Methods in Enzymology

Volume 221

Membrane Fusion Techniques

Part B

EDITED BY

Nejat Düzgüneş

DEPARTMENT OF MICROBIOLOGY
UNIVERSITY OF THE PACIFIC SCHOOL OF DENTISTRY
SAN FRANCISCO, CALIFORNIA

D1401991

ACADEMIC PRESS, INC.
A Division of Harcourt Brace & Company
San Diego   New York   Boston   London   Sydney   Tokyo   Toronto

"DISCARDED"
Magale Library
Southern Arkansas University
Magnolia, Arkansas

Academic Press, Inc.
1250 Sixth Avenue, San Diego, California 92101-4311

United Kingdom Edition published by
Academic Press Limited
24–28 Oval Road, London NW1 7DX

International Standard Serial Number: 0076-6879

International Standard Book Number: 0-12-182122-6

PRINTED IN THE UNITED STATES OF AMERICA
93 94 95 96 97 98   EB   9 8 7 6 5 4 3 2 1

Table of Contents

Section IV. Intracellular Membrane Fusion

Section V. Membrane Fusion in Fertilization

Section VI. Introduction of Macromolecules into Cells by Membrane Fusion

Section VII. Protoplast Fusion

Contributors to Volume 221

Article numbers are in parentheses following the names of contributors.
Affiliations listed are current.

GUDRUN AHNERT-HILGER (11), *Abteilung Gastroenterologie, Klinikum Steglitz, Freien Universität Berlin, W-1000 Berlin, Germany*

PER ASHORN (2), *Laboratory of Immunopathology, Department of Biochemical Sciences, University of Tampere, Tampere, Finland*

W. E. BALCH (17), *Department of Molecular Biology, Research Institute of Scripps Clinic, La Jolla, California 92037*

C. J. M. BECKERS (17), *Department of Molecular Biology, Research Institute of Scripps Clinic, La Jolla, California 92037*

EDWARD A. BERGER (2), *Laboratory of Viral Disease, National Institute of Allergy and Infectious Diseases, National Institutes of Health, Bethesda, Maryland 20892*

CORDIAN BEYER (11), *MRC, Neuroendocrine Group, Institute of Animal Physiology, Brabhan, Cambridge CB2 4AT, England*

LEA K. BLEYMAN (14), *Department of Natural Sciences, Baruch College, State University of New York, New York, New York 10010*

ROBERT BLUMENTHAL (4), *Section of Membrane Structure and Function, Laboratory of Mathematical Biology, National Cancer Institute, National Institutes of Health, Bethesda, Maryland 20892*

DOUGLAS E. CHANDLER (9), *Department of Zoology, Arizona State University, Tempe, Arizona 85287*

MARIA I. COLOMBO (16), *Department of Cell Biology and Physiology, Washington University School of Medicine, St. Louis, Missouri 63110*

CARL E. CREUTZ (15), *Department of Pharmacology, University of Virginia, Charlottesville, Virginia 22908*

M. R. DAVEY (29), *Plant Genetic Manipulation Group, Department of Life Science, University of Nottingham, University Park, Nottingham NG7 2RD, England*

H. DAVIDSON (17), *Department of Molecular Biology, Research Institute of Scripps Clinic, La Jolla, California 92037*

RUBEN DIAZ (16), *Department of Pediatrics, Children's Hospital, Boston, Massachusetts 02115*

ROBERT W. DOMS (5), *Department of Pathology and Laboratory Medicine, Philadelphia, Pennsylvania 19104*

ARNOLD J. M. DRIESSEN (30), *Department of Microbiology, University of Groningen, 9721 NN Haren, The Netherlands*

NEJAT DÜZGÜNEŞ (7, 18, 23), *Department of Microbiology, University of the Pacific School of Dentistry, San Francisco, California 94115, and Department of Pharmaceutical Chemistry, University of California, San Francisco, San Francisco, California 94143*

PHILIP L. FELGNER (23), *Vical, Inc., San Diego, California 92121*

KARL J. FÖHR (12), *Naturwissenschaftliches und Medizinisches Institut, Universität Tübingen in Reutlingen, D-7410 Reutlingen, Germany*

JEFFREY S. GLENN (26), *Department of Pharmacology, University of California, San Francisco, San Francisco, California 94143*

MAYER B. GOREN (18), *Department of Molecular and Cellular Biology, National Jewish Center for Immunology and Respiratory Medicine, Denver, Colorado 80206*

MANFRED GRATZL (11, 12), *Abteilung Anatomie und Zellbiologie, Universität Ulm, Oberer Eselsberg, D-7900 Ulm, Germany*

MARISA JACONI (13), *Division of Infectious Diseases, University Hospital of Geneva, CH-1211 Geneva 4, Switzerland*

RAYMOND T. KADO (22), *Centre National de la Recherche Scientifique, Laboratoire de Neurobiologie Cellulaire et Moléculaire, 91198 Gif-sur-Yvette Cedex, France*

YASUFUMI KANEDA (25), *Institute for Molecular and Cellular Biology, Osaka University, Suita, Osaka 565, Japan*

KEIKO KATO (25), *Institute for Molecular and Cellular Biology, Osaka University, Suita, Osaka 565, Japan*

DEREK E. KNIGHT (10), *Division of Biomedical Science, King's College London, London WC2R 2LS, England*

WIL N. KONINGS (30), *Department of Microbiology, University of Groningen, 9751 NN Haren, The Netherlands*

JAMES M. LENHARD (16), *Department of Cell Biology and Physiology, Washington University School of Medicine, St. Louis, Missouri 63110*

W. J. LENNARZ (21), *Department of Biochemistry and Cell Biology, State University of New York at Stony Brook, Stony Brook, New York 11794*

DANIEL P. LEW (13), *Division of Infectious Diseases, University Hospital of Geneva, CH-1211 Geneva 4, Switzerland*

JEFFREY D. LIFSON (1), *Division of Cellular Immunology, Genelabs Incorporated, Redwood City, California 94063*

FRANK J. LONGO (20), *Department of Anatomy, University of Iowa, Iowa City, Iowa 52242*

PAUL F. LURQUIN (31), *Department of Genetics and Cell Biology, Washington State University, Pullman, Washington 99164*

P. T. LYNCH (29), *Plant Genetic Manipulation Group, Department of Life Science, University of Nottingham, University Park, Nottingham NG7 2RD, England*

SADHANA MAJUMDAR (18), *Institute of Microbial Technology, Chandigarh 160014, India*

LUIS S. MAYORGA (16), *Instituto de Histologia y Embriologia, Facultad de Ciencias Medicas, Universidad Nacional de Cuyo (CONICET), Mendoza 5500, Argentina*

CARRIE J. MERKLE (9), *Department of Physiology, College of Medicine, University of Arizona, Tuscon, Arizona 85712*

STEPHEN J. MORRIS (4), *Division of Molecular Biology and Biochemistry, School of Biological Sciences, University of Missouri-Kansas City, Kansas City, Missouri 64110*

BERNARD MOSS (2), *Laboratory of Viral Diseases, National Institute of Allergy and Infectious Diseases, National Institutes of Health, Bethesda, Maryland 20892*

MAHITO NAKANISHI (25), *Institute for Molecular and Cellular Biology, Osaka University, Suita, Osaka 565, Japan*

G. A. NEIL (27), *Department of Internal Medicine, University of Iowa, Iowa City, Iowa 52242*

YOSHIO OKADA (3), *Institute for Molecular and Cellular Biology, Osaka University, Suita, Osaka 565, Japan*

S. PIND (17), *Department of Molecular Biology, Research Institute of Scripps Clinic, La Jolla, California 92037*

H. PLUTNER (17), *Department of Molecular Biology, Research Institute of Scripps Clinic, La Jolla, California 92037*

J. B. POWER (29), *Plant Genetic Manipulation Group, Department of Life Science, University of Nottingham, University Park, Nottingham NG7 2RD, England*

TULLIO POZZAN (13), *Università di Padova, Instituto di Patologia Generale, Padova, Italy*

FRANCO ROLLO (31), *Università degli Studi di Camerino, Dipartimento di Biologia Cellulare, 62032 Camerino, Italy*

N. RUIZ-BRAVO (21), *Genetics Program, National Institute of General Medical Sciences, Bethesda, Maryland 20892*

DEBI P. SARKAR (4), *Department of Biochemistry, University of Delhi, South Campus, New Delhi 110021, India*

BIRGIT H. SATIR (14), *Department of Anatomy and Structural Biology, Albert Einstein College of Medicine, Bronx, New York 10461*

S. L. SCHMID (17), *Department of Molecular Biology, Research Institute of Scripps Clinic, La Jolla, California 92037*

R. SCHWANINGER (17), *Department of Molecular Biology, Research Institute of Scripps Clinic, La Jolla, California 92037*

MICHAEL C. SCRUTTON (10), *Division of Life Science, King's College London, London WC2R 2LS, England*

PHILIP D. STAHL (16), *Department of Cell Biology and Physiology, Washington University School of Medicine, St. Louis, Missouri 63110*

BRIGITTE STECHER (11), *Lehrstuhl für Entwicklungsneurobiologie, Fakultät für Biologie, Universität Konstanz, D-7750 Konstanz, Germany*

ROBERT M. STRAUBINGER (28), *Department of Pharmaceutics, School of Pharmacy,* *State University of New York at Buffalo, Amherst, New York 14260*

TSUYOSHI UCHIDA[1] (25), *Institute for Molecular and Cellular Biology, Osaka University, Suita, Osaka 565, Japan*

PAUL S. USTER (19), *Liposome Technology, Inc., Menlo Park, California 94025*

WOJCIECH WARCHOL (12), *Abteilung Anatomie und Zellbiologie, Universität Ulm, Oberer Eselsberg, D-7900 Ulm, Germany*

JUDITH M. WHITE (26), *Department of Pharmacology, University of California, San Francisco, San Francisco, California 94143*

RYUZO YANAGIMACHI (20), *Department of Anatomy and Reproductive Biology, University of Hawaii Medical School, Honolulu, Hawaii 96822*

YOSHIHIRO YONEDA (24), *Department of Anatomy, Osaka University Medical School, Suita, Osaka 565, Japan*

TETSURO YOSHIMURA (6), *Institute for Enzyme Research, University of Tokushima, Tokushima 770, Japan*

JOSHUA ZIMMERBERG (4, 8), *Laboratory of Theoretical and Physical Biology, National Institute of Child Health and Development, National Institutes of Health, Bethesda, Maryland 20892*

ULRICH ZIMMERMANN (27), *Lehrstuhl für Biotechnologie, Universität Würzburg, D-8700 Würzburg, Germany*

[1] Deceased.

Preface

To commemorate the twenty-first anniversary of the publication of J. D. Watson and F. H. C. Crick's famous article on the structure of DNA, the April 26, 1974, issue of *Nature* featured a special section entitled "Molecular biology comes of age." While the origin of the field of membrane fusion research cannot be traced to a single article, two comprehensive reviews on virus-induced cell fusion and on membrane fusion appeared in 1972 and 1973, respectively (G. Poste, *Int. Rev. Cytol.* **33,** 157–252; G. Poste and A. C. Allison, *Biochim. Biophys. Acta* **300,** 421–465). In the two decades since, there has been a rapid growth in the number of studies on the molecular mechanisms of membrane fusion, culminating in several books on the subject (A. E. Sowers, ed., "Cell Fusion," Plenum Press, 1987; S. Ohki, D. Doyle, T. D. Flanagan, S. W. Hui, and E. Mayhew, eds., "Molecular Mechanisms of Membrane Fusion," Plenum Press, 1988; N. Düzgüneş, ed., "Membrane Fusion in Fertilization, Cellular Transport, and Viral Infection," Academic Press, 1988; J. Wilschut and D. Hoekstra, eds., "Membrane Fusion," Marcel Dekker, 1991). With the publication of Volumes 220 and 221 of *Methods in Enzymology* dedicated to this subject, it is not entirely inappropriate to declare the field of membrane fusion as having come of age.

The chapters in this and the accompanying Volume 220 present not only the details of methods used in membrane fusion research, but also a critical analysis of the methods, their advantages and shortcomings, and possible artifacts. While several sections focus on the elucidation of the mechanisms of fusion in various experimental systems (Fusion of Liposomes and Other Artificial Membranes; Fusion of Viruses with Target Membranes; Cell–Cell Fusion Mediated by Viruses and Viral Proteins; Conformational Changes of Proteins during Membrane Fusion; Membrane Fusion during Exocytosis; Intracellular Membrane Fusion; Membrane Fusion in Fertilization), several others describe applications of membrane fusion technology (Induction of Cell–Cell Fusion; Introduction of Macromolecules into Cells by Membrane Fusion; Protoplast Fusion). The methodology presented should be of value not only to newcomers to membrane fusion research who wish to employ some of the techniques described in these books, but also to researchers in the field who need to adopt an alternative technique.

I would like to thank the contributors to this volume, without whose willing and able collaboration this work would not even have begun. I would also like to express my appreciation for their patience with me and with their fellow authors, not all of whom were able to submit their

manuscripts at the same time. I thank Shirley Light of Academic Press for
her patience, understanding, encouragement, and persistence in producing
this volume, and Cynthia Vincent for her invaluable editorial assistance. I
also thank my wife Diana Flasher for her constant support and enthusiasm
for this project, despite countless weekends I spent editing manuscripts.
Finally, I wish to dedicate this volume to my aunt Sevim Uygurer, my
brother Arda Düzgüneş, and my wife Diana Flasher, in grateful appreciation of their love, support, and understanding.

NEJAT DÜZGÜNEŞ

METHODS IN ENZYMOLOGY

VOLUME XVII. Metabolism of Amino Acids and Amines (Parts A and B)
Edited by HERBERT TABOR AND CELIA WHITE TABOR

VOLUME XVIII. Vitamins and Coenzymes (Parts A, B, and C)
Edited by DONALD B. MCCORMICK AND LEMUEL D. WRIGHT

VOLUME XIX. Proteolytic Enzymes
Edited by GERTRUDE E. PERLMANN AND LASZLO LORAND

VOLUME XX. Nucleic Acids and Protein Synthesis (Part C)
Edited by KIVIE MOLDAVE AND LAWRENCE GROSSMAN

VOLUME XXI. Nucleic Acids (Part D)
Edited by LAWRENCE GROSSMAN AND KIVIE MOLDAVE

VOLUME XXII. Enzyme Purification and Related Techniques
Edited by WILLIAM B. JAKOBY

VOLUME XXIII. Photosynthesis (Part A)
Edited by ANTHONY SAN PIETRO

VOLUME XXIV. Photosynthesis and Nitrogen Fixation (Part B)
Edited by ANTHONY SAN PIETRO

VOLUME XXV. Enzyme Structure (Part B)
Edited by C. H. W. HIRS AND SERGE N. TIMASHEFF

VOLUME XXVI. Enzyme Structure (Part C)
Edited by C. H. W. HIRS AND SERGE N. TIMASHEFF

VOLUME XXVII. Enzyme Structure (Part D)
Edited by C. H. W. HIRS AND SERGE N. TIMASHEFF

VOLUME XXVIII. Complex Carbohydrates (Part B)
Edited by VICTOR GINSBURG

VOLUME XXIX. Nucleic Acids and Protein Synthesis (Part E)
Edited by LAWRENCE GROSSMAN AND KIVIE MOLDAVE

VOLUME XXX. Nucleic Acids and Protein Synthesis (Part F)
Edited by KIVIE MOLDAVE AND LAWRENCE GROSSMAN

VOLUME XXXI. Biomembranes (Part A)
Edited by SIDNEY FLEISCHER AND LESTER PACKER

VOLUME XXXII. Biomembranes (Part B)
Edited by SIDNEY FLEISCHER AND LESTER PACKER

VOLUME XXXIII. Cumulative Subject Index Volumes I–XXX
Edited by MARTHA G. DENNIS AND EDWARD A. DENNIS

VOLUME XXXIV. Affinity Techniques (Enzyme Purification: Part B)
Edited by WILLIAM B. JAKOBY AND MEIR WILCHEK

VOLUME XXXV. Lipids (Part B)
Edited by JOHN M. LOWENSTEIN

VOLUME LV. Biomembranes (Part F: Bioenergetics)
Edited by SIDNEY FLEISCHER AND LESTER PACKER

VOLUME LVI. Biomembranes (Part G: Bioenergetics)
Edited by SIDNEY FLEISCHER AND LESTER PACKER

VOLUME LVII. Bioluminescence and Chemiluminescence
Edited by MARLENE A. DeLUCA

VOLUME LVIII. Cell Culture
Edited by WILLIAM B. JAKOBY AND IRA PASTAN

VOLUME LIX. Nucleic Acids and Protein Synthesis (Part G)
Edited by KIVIE MOLDAVE AND LAWRENCE GROSSMAN

VOLUME LX. Nucleic Acids and Protein Synthesis (Part H)
Edited by KIVIE MOLDAVE AND LAWRENCE GROSSMAN

VOLUME 61. Enzyme Structure (Part H)
Edited by C. H. W. HIRS AND SERGE N. TIMASHEFF

VOLUME 62. Vitamins and Coenzymes (Part D)
Edited by DONALD B. McCORMICK AND LEMUEL D. WRIGHT

VOLUME 63. Enzyme Kinetics and Mechanism (Part A: Initial Rate and Inhibitor Methods)
Edited by DANIEL L. PURICH

VOLUME 64. Enzyme Kinetics and Mechanism (Part B: Isotopic Probes and Complex Enzyme Systems)
Edited by DANIEL L. PURICH

VOLUME 65. Nucleic Acids (Part I)
Edited by LAWRENCE GROSSMAN AND KIVIE MOLDAVE

VOLUME 66. Vitamins and Coenzymes (Part E)
Edited by DONALD B. McCORMICK AND LEMUEL D. WRIGHT

VOLUME 67. Vitamins and Coenzymes (Part F)
Edited by DONALD B. McCORMICK AND LEMUEL D. WRIGHT

VOLUME 68. Recombinant DNA
Edited by RAY WU

VOLUME 69. Photosynthesis and Nitrogen Fixation (Part C)
Edited by ANTHONY SAN PIETRO

VOLUME 70. Immunochemical Techniques (Part A)
Edited by HELEN VAN VUNAKIS AND JOHN J. LANGONE

VOLUME 71. Lipids (Part C)
Edited by JOHN M. LOWENSTEIN

VOLUME 72. Lipids (Part D)
Edited by JOHN M. LOWENSTEIN

VOLUME 213. Carotenoids (Part A: Chemistry, Separation, Quantitation, and Antioxidation)
Edited by LESTER PACKER

VOLUME 214. Carotenoids (Part B: Metabolism, Genetics, and Biosynthesis)
Edited by LESTER PACKER

VOLUME 215. Platelets: Receptors, Adhesion, Secretion (Part B)
Edited by JACEK J. HAWIGER

VOLUME 216. Recombinant DNA (Part G)
Edited by RAY WU

VOLUME 217. Recombinant DNA (Part H)
Edited by RAY WU

VOLUME 218. Recombinant DNA (Part I)
Edited by RAY WU

VOLUME 219. Reconstitution of Intracellular Transport
Edited by JAMES E. ROTHMAN

VOLUME 220. Membrane Fusion Techniques (Part A)
Edited by NEJAT DÜZGÜNEŞ

VOLUME 221. Membrane Fusion Techniques (Part B)
Edited by NEJAT DÜZGÜNEŞ

VOLUME 222. Proteolytic Enzymes in Coagulation, Fibrinolysis, and Complement Activation (Part A: Mammalian Blood Coagulation Factors and Inhibitors) (in preparation)
Edited by LASZLO LORAND AND KENNETH G. MANN

VOLUME 223. Proteolytic Enzymes in Coagulation, Fibrinolysis, and Complement Activation (Part B: Complement Activation, Fibrinolysis, and Nonmammalian Blood Coagulation Factors) (in preparation)
Edited by LASZLO LORAND AND KENNETH G. MANN

VOLUME 224. Molecular Evolution: Producing the Biochemical Data (in preparation)
Edited by ELIZABETH ANNE ZIMMER, THOMAS J. WHITE, REBECCA L. CANN, AND ALLAN C. WILSON

VOLUME 225. Guide to Techniques in Mouse Development (in preparation)
Edited by PAUL M. WASSARMAN AND MELVIN L. DEPAMPHILIS

VOLUME 226. Metallobiochemistry (Part C: Spectroscopic and Physical Methods for Probing Metal Ion Environments in Metalloenzymes and Metalloproteins) (in preparation)
Edited by JAMES F. RIORDAN AND BERT L. VALLEE

VOLUME 227. Metallobiochemistry (Part D: Physical and Spectroscopic Methods for Probing Metal Ion Environments in Metalloproteins) (in preparation)
Edited by JAMES F. RIORDAN AND BERT L. VALLEE

Section I

Cell–Cell Fusion Mediated by Viruses and Viral Proteins

[1] Fusion of Human Immunodeficiency Virus-Infected Cells with Uninfected Cells

By Jeffrey D. Lifson

Introduction

Processes involving membrane fusion events are an essential part of the life cycle of many pathogenic enveloped viruses.[1] Fusion between the virion envelope and either the plasmalemma of target cells or the membrane of a cellular endocytic vacuole may be required for establishment of productive infection.[1-3] In addition, virally induced cell fusion gives rise to the multinucleated giant cells, or *syncytia,* typically observed in many *in vitro* assay systems used to study replication of these viruses, and in histological sections of tissues infected *in vivo.*[4]

Perhaps the best characterized of these viruses, with regard to mechanistic aspects of virally induced membrane fusion phenomena, is the orthomyxovirus influenzavirus.[3,5-9] In the influenzavirus system, the envelope protein (hemagglutinin, HA) precursor HA_0 is synthesized in infected cells and cleaved to yield two subunits (designated HA_1 and HA_2), which are present on the plasma membrane of infected cells as well as on the lipid bilayer of budded viral particles.[5,10,11] The endoproteolytic cleavage that generates HA_1 and HA_2 also exposes a highly conserved hydrophobic stretch (approximately 30 residues) within HA_0, which becomes the amino-terminal domain of HA_2. It has been postulated that binding of the trimeric form of HA produces a conformational change that positions this hydrophobic domain in proximity to host cell membranes.[9] Entry of this hydrophobic domain of HA_2 into the lipid bilayer of a host cell membrane is believed to initiate membrane fusion, which results in fusion of the

[1] J. White, M. Kielian, and A. Helenius, *Q. Rev. Biophys.* **16**, 151 (1983).

[2] A. Scheid and P. W. Choppin, *Virology* **57**, 475 (1974).

[3] K. S. Matlin, H. Reggio, A. Helenius, and K. Simons, *J. Cell Biol.* **91**, 601 (1981).

[4] G. Poste and C. A. Pasternak, *Cell Surf. Rev.* **5**, 305 (1978).

[5] S. Lazarowitz and P. W. Choppin, *Virology* **68**, 440 (1975).

[6] C. D. Richardson, A. Scheid, and P. W. Choppin, *Virology* **105**, 205 (1980).

[7] J. Skehel, P. Bayley, E. Brown, S. Martin, M. Waterfield, J. White, I. Wilson, and D. Wiley, *Proc. Natl. Acad. Sci. U.S.A.* **79**, 968 (1982).

[8] J. White, A. Helenius, and M. Gething, *Nature (London)* **300**, 658 (1982).

[9] I. A. Wilson, J. J. Skehel, and D. C. Wiley, *Nature (London)* **289**, 366 (1981).

[10] S. Lazarowitz, R. W. Compans, and P. W. Choppin, *Virology* **68**, 199 (1973).

[11] H.-D. Klenk, R. Rott, M. Orlich, and J. Blodorn, *Virology* **68**, 426 (1975).

virion envelope and the host membrane, or fusion of the plasma membranes of infected and uninfected cells.[1]

In broad outline, the influenzavirus system appears to represent a generalizable model of virally induced membrane fusion events. In this general scheme, heteromultimeric viral envelope glycoproteins are generated from polyprotein precursors by an endoproteolytic cleavage, specified by the host cell. This cleavage liberates a highly conserved, hydrophobic domain within the precursor protein that forms the amino-terminal portion of one of the resulting proteins; the conformation of the resulting envelope glycoprotein complex presumably sequesters this hydrophobic amino-terminal domain away from the aqueous environment at the surface of the cell membrane or viral particle. The second component of the viral envelope glycoprotein, associated with the first, interacts specifically with a receptor moiety on susceptible target cells, in part defining the host range of the virus. In conjunction with binding of this targeting portion of the envelope glycoprotein to a receptor moiety, a conformational change is believed to expose the hydrophobic fusion domain within the viral envelope glycoprotein, initiating membrane fusion events. This model appears to apply to both orthomyxoviruses and paramyxoviruses.[1,2,6,12–14]

This general model of virus-induced membrane fusion also appears to apply to human immunodeficiency virus[15] (HIV), the etiological agent for the acquired immunodeficiency syndrome (AIDS). The envelope glycoprotein of HIV is synthesized as a 160-kDa precursor (gp160), then cleaved by a host cell-specified protease into an external glycoprotein of 120 kDa (gp120) that is noncovalently associated with a transmembrane envelope glycoprotein of 41 kDa (gp41).[16] gp120 mediates specific binding to the cellular glycoprotein known as CD4,[17,18] which serves as a specific receptor for HIV,[19–21] in addition to its role in the regulation of immune re-

[12] M.-C. Hsu, A. Scheid, and P. Choppin, *Virology* **95**, 476 (1979).
[13] M.-C. Hsu, A. Scheid, and P. W. Choppin, *J. Biol. Chem.* **256**, 3557 (1981).
[14] A. Scheid and P. W. Choppin, *Virology* **69**, 265 (1976).
[15] In this chapter the term HIV, when used without additional qualification, refers to human immunodeficiency virus, type 1.
[16] J. M. McCune, L. B. Rabin, M. B. Feinberg, M. Lieberman, J. C. Kosek, G. R. Reyes, and I. L. Weissman, *Cell (Cambridge, Mass.)* **53**, 55 (1988).
[17] J. S. McDougal, M. S. Kennedy, J. M. Sligh, S. P. Cort, A. Mawle, and J. K. A. Nicholson, *Science* **231**, 382 (1986).
[18] L. A. Lasky, G. Nakamura, D. H. Smith, C. Fennie, C. Shimasaki, E. Patzer, P. Berman, T. Gregory, and D. J. Capon, *Cell (Cambridge, Mass.)* **50**, 975 (1987).
[19] A. G. Dalgleish, P. C. L. Beverly, P. R. Clapham, D. H. Crawford, M. F. Greaves, and R. A. Weiss, *Nature (London)* **312**, 763 (1984).
[20] D. Klatzmann, E. Champagne, S. Chamaret, J. Gruest, D. Guétard, T. Hercend, J.-C. Gluckman, and L. Montagnier, *Nature (London)* **312**, 767 (1984).
[21] P. J. Maddon, A. G. Dalgleish, J. S. McDougal, P. R. Clapham, R. A. Weiss, and R. Axel, *Cell (Cambridge, Mass.)* **47**, 333 (1986).

sponses.[22] gp41, which contains a hydrophobic amino-terminal domain, represents the fusion-mediating component of the HIV envelope glycoprotein complex.[16,23] Interactions between gp120 and CD4, leading to gp41-dependent membrane fusion events, are necessary to mediate virion infectivity and cell–cell fusion in susceptible target cells.[23-26] The binding and fusion events involved in virion infectivity[27] and fusion of infected, HIV envelope-expressing cells with uninfected CD4-expressing cells appear to be largely similar, although higher concentrations of blocking agents are generally required to inhibit cell fusion than to inhibit virion infectivity, perhaps owing to the higher valency of the molecular interactions involved. *In vitro* studies have suggested that there may be subtle differences between the binding and membrane fusion events involved in virion infectivity and HIV-induced cell–cell fusion, although the structural and mechanistic basis of such apparent differences remains obscure.[28]

These binding interactions and resultant fusion events represent obligate aspects of the life cycle of HIV and are believed to be among the primary determinants of viral cytopathogenicity.[23-26] The obligate role of these processes in the life cycle of HIV and in the ability of HIV to kill susceptible cells has led to efforts to use CD4-based molecules as a form of therapeutic intervention in HIV infection, although results from clinical studies have been disappointing.[29-35] The importance of CD4-dependent

[22] J. D. Lifson and E. G. Engleman, *Immunol. Rev.* **109**, 93 (1989).
[23] M. Kowalski, J. Potz, L. Basiripour, T. Dorfman, W. C. Goh, E. Terwilliger, A. Dayton, C. Rosen, W. Haseltine, and J. Sodroski, *Science* **237**, 1351 (1987).
[24] J. D. Lifson, G. R. Reyes, M. S. McGrath, B. S. Stein, and E. G. Engleman, *Science* **232**, 1123 (1986).
[25] J. D. Lifson, M. B. Feinberg, G. R. Reyes, L. Rabin, B. Banapour, S. Chakrabarti, B. Moss, F. Wong-Staal, K. S. Steimer, and E. G. Engleman, *Nature (London)* **323**, 725 (1986).
[26] J. Sodroski, W. C. Goh, C. Rosen, K. Campbell, and W. A. Haseltine, *Nature (London)* **322**, 470 (1986).
[27] B. S. Stein, S. D. Gowda, J. D. Lifson, R. C. Penhallow, K. G. Bensch, and E. G. Engleman, *Cell (Cambridge, Mass.)* **49**, 659 (1987).
[28] L. E. Eiden and J. D. Lifson, *Immunol. Today* **13**, 201 (1992).
[29] J. D. Lifson, K. M. Hwang, P. L. Nara, B. Fraser, M. Padgett, N. M. Dunlop, and L. E. Eiden, *Science* **241**, 712 (1988).
[30] P. L. Nara, K. M. Hwang, D. M. Rausch, J. D. Lifson, and L. E. Eiden, *Proc. Natl. Acad. Sci. U.S.A.* **86**, 7139 (1989).
[31] R. A. Fisher, J. M. Bertonis, W. Meier, V. A. Johnson, D. S. Costopoulos, T. Liu, R. Tizard, B. D. Walker, M. S. Hirsch, R. T. Schooley, and R. A. Flavell, *Nature (London)* **331**, 76 (1988).
[32] R. E. Hussey, N. E. Richardson, M. Kowalski, N. R. Brown, H.-C. Chang, R. F. Siliciano, T. Dorfman, B. Walker, J. Sodroski, and E. L. Reinherz *Nature (London)* **331**, 78 (1988).
[33] K. C. Deen, J. S. McDougal, R. Inacker, G. Folena-Wasserman, J. Arthos, J. Rosenberg, P. J. Maddon, R. Axel, and R. W. Sweet, *Nature (London)* **331**, 82 (1988).
[34] A. Traunecker, W. Luke, and K. Karjalainen, *Nature (London)* **331**, 84 (1988).
[35] D. H. Smith, R. A. Byrn, S. A. Marsters, T. Gregory, J. E. Groopman, and D. J. Capon, *Science* **238**, 1704 (1987).

HIV envelope-induced cell fusion makes it highly desirable to have a simple, convenient, and rapid assay to study this process. This chapter describes such an assay, which has been utilized in our laboratory with various modifications in both analytical studies of HIV-induced cell fusion[24,25] and in efforts to identify specific inhibitors of this process.[22,28,29]

Methods

Biosafety

The assay method described below uses HIV-infected cells, which pose an infectious hazard. Although infection with HIV in the laboratory setting is highly unlikely if appropriate procedures are followed and an appropriate facility and equipment utilized, infection of laboratory workers with HIV, apparently through occupational exposure, has been documented.[36,37] Readers are referred elsewhere for detailed descriptions of the relevant facilities and procedures required for handling infectious HIV[38] and reminded that laboratory studies with this pathogen should be performed only by appropriately trained and qualified personnel, subject to appropriate institutional monitoring.

CD4-Dependent HIV-Induced Cell Fusion: Assay Method

Overview. The basic principle of the assay described below is straightforward. It involves the cocultivation of HIV-infected, HIV envelope glycoprotein-expressing cells with uninfected cells expressing CD4. Multinucleated giant cells (syncytia) arising from HIV envelope glycoprotein-induced, CD4-dependent cell fusion events are evaluated by inverted phase-contrast microscopy and scored on a qualitative or quantitative basis. A particular version of this assay approach is detailed below. This general approach can be adapted to different cell lines, viral isolates, time period, and readout methods as dictated by the experimental questions of interest.

Cell Lines. The CD4+ T lymphoblastic cell line VB is employed in our laboratory as an indicator cell in fusion assays. This cell line is notable for its high level of cell surface expression of CD4 and consequent

[36] S. H. Weiss, J. J. Goedert, S. Gartner, M. Popovič, D. Waters, P. Markham, F. Di Marzo Veronese, M. H. Gail, W. E. Barkley, J. Gibbons, F. A. Gill, M. Leuther, G. M. Shaw, R. C. Gallo, and W. A. Blattner, *Science* **239**, 68 (1988).

[37] *Morbid. Mortal. Wkly. Rep.* **37**, 19 (1988).

[38] *Morbid. Mortal. Wkly. Rep.* **37**, 1 (1988).

extreme susceptibility to HIV infection and HIV envelope-induced cell fusion.[22,25,27] Cells are propagated in RPMI 1640 supplemented with 10% (v/v) heat-inactivated fetal bovine serum (HI-FBS) and 2 mM L-glutamine by passage twice weekly. Cells should be in logarithmic growth for use in assay procedures. The VB cell line is available through the National Institutes of Health (NIH)-sponsored AIDS Research and Reference Reagent Program Repository (c/o Ogden BioServices Corporation, 685 Lofstrand Land, Rockville, MD 20850).

Cells of the CD4[+] T lymphoblastoid cell line H9[39] chronically infected with HIV serve as the HIV envelope glycoprotein-expressing component in the fusion assay. In the configuration most frequently used in our laboratory, cells chronically infected with HIV-1$_{HXB-2}$[40] are used as the infected cell partner. These chronically infected cells are propagated in RPMI 1640 supplemented with 10% (v/v) HI-FBS and 2 mM L-glutamine by passage twice weekly. Serial immunohistofluorescence analysis for HIV antigens confirmed stable expression of HIV proteins by these cells over several months of continuous passage, with ≥95% of these cells productively infected. Cells should be in logarithmic growth for use in assay procedures. Although this chapter refers to H9 cells chronically infected with HIV-1$_{HXB-2}$, we have also worked with H9 cells chronically infected with other HIV-1 isolates as well. It is important to note the extent of HIV expression in the chronically infected H9 population used, and to perform control experiments to characterize the extent and kinetics of syncytium formation when other infected cell populations are used. However, with these caveats, the assay procedure described here can be used to study any syncytium-inducing strain of HIV. Indeed, comparative studies of well-characterized, distinct HIV isolates with documented differences in the sequences of their envelope genes have provided insight into the conserved and variable regions involved in neutralization of HIV[41] and are an essential component in the evaluation of compounds intended as inhibitors of HIV-induced binding/fusion events. H9 cells and H9 cells chronically infected with a variety of characterized HIV-1 isolates are available through the AIDS Research and Reference Reagent Program Repository.

One of the reasons for using H9 cells as the infected cell component in the fusion assay is that H9 cells express a relatively low level of cell surface CD4. The level expressed is sufficient to render the cells susceptible to HIV

---

[39] M. Popovič, M. G. Sarngadharan, E. Read, and R. C. Gallo, *Science* **224**, 497 (1984).
[40] M. Robert-Guroff, M. S. Reitz, W. G. Robey, and R. C. Gallo, *J. Immunol.* **137**, 3306 (1987).
[41] R. A. Weiss, P. R. Clapham, J. N. Weber, A. G. Dalgleish, L. A. Lasky, and P. W. Berman, *Nature (London)* **324**, 572 (1986).

infection, but low enough that not all cells are killed by CD4-dependent viral cytopathic effects following infection.[27,39] In addition, following HIV infection, cell surface expression of CD4 is down regulated through mechanisms involving formation of intracytoplasmic gp120–CD4 complexes and, most likely, through other processes as well.[42] The net result of these phenomena is that, following acute infection of H9 cell cultures with HIV, a chronically infected outgrowth can be isolated that is productively infected, displaying HIV envelope glycoproteins on cell membranes, but lacking detectable cell surface expression of CD4. Consequently, these CD4⁻ cells do not fuse with one another, but are capable of readily inducing syncytia when cocultivated with CD4-expressing uninfected cells.[43]

Assay Procedure. Although the assay configuration described below is intended to evaluate putative inhibitors of HIV envelope/CD4 binding/fusion events, the procedures can be readily modified to suit other applications. For compounds believed to inhibit on the basis of specific interactions with infected cells, the order of incubation should be as described below. For compounds believed to inhibit on the basis of specific interactions with uninfected CD4⁺ cells, the order should be reversed.

To assay inhibitors, serial dilutions in phosphate-buffered saline (PBS) are prepared using a multichannel pipettor and a round-bottom 96-well microtiter plate. Adding 100 μl of PBS to 100 μl of inhibitor conveniently provides 12 columns per plate representing 2-fold dilution series from $1:2^1$ to $1:2^8$, a range sufficient to allow end-point titration of most test compounds. Replicate dilution series to assess reproducibility of results can be performed in adjacent columns on the plate. Phosphate-buffered saline alone serves as a negative control whereas either a soluble CD4 preparation or a suitable CD4-reactive murine monoclonal antibody (such as anti-Leu 3a or OKT4a), known to block HIV-induced syncytium induction, serves as a positive control for inhibition of cell fusion.

Chronically infected H9 cells in logarithmic phase growth are resuspended to 2×10^6 in RPMI 1640 supplemented with 10% (v/v) HI-FBS and 2 mM L-glutamine and plated in the wells of flat-bottom 96-well microtiter plates at 5×10^4 cells in 25 μl/well. Using a multichannel pipettor, aliquots (50 μl/well) of appropriate dilutions of test compounds are transferred with well-to-well concordance from the plate in which serial dilutions of test compounds had been prepared previously. Infected cells are incubated with compounds for 30–60 min at 37° in a humidified

[42] J. A. Hoxie, J. D. Alpers, J. L. Rackowski, K. Huebner, B. S. Haggarty, A. J. Cedarbaum, and J. C. Reed, *Science* **234**, 1123 (1986).

[43] J. Lifson, S. Coutre, E. Huang, and E. Engleman, *J. Exp. Med.* **164**, 2101 (1986).

incubator with a 5% CO_2/95% air atmosphere. After incubation of infected cells with test compounds for the desired period of time, VB cells are added. VB cells in logarithmic phase growth are resuspended to 2×10^6 in RPMI 1640 supplemented with 10% (v/v) HI-FBS and 2 mM L-glutamine and 5×10^4 cells in 25 μl is then added to each well containing infected H9 cells and test compound. Cells are then cocultivated in the continuous presence of test compounds at 37° in a humidified incubator with a 5% CO_2/95% air atmosphere for the desired period. In untreated wells, syncytia generally begin to form within 1–3 hr after mixing of the infected and uninfected cell populations. Presyncytial clusters are first apparent as clumps of uninfected VB cells surrounding central infected H9 cells. Syncytia are typically well developed by 4–6 hr after mixing and are readily identifiable as large multicellular bodies surrounded by rosette-like clusters of uninfected VB cells. Syncytia continue to increase in number and extent over the subsequent 8–20 hr, becoming quite large, with some syncytia incorporating several hundred cells (Fig. 1). After 24 hr postmixing, no new syncytia typically form, but the previously formed syncytia undergo degenerative changes, characterized by nuclear pyknosis and cytoplasmic ballooning as the cells die. Syncytia are thus best scored at some time between 4 and 24 hr after initiation of cocultivation.

Scoring of Assay. The assay is evaluated by inverted phase-contrast microscopy, typically at $\times 100$ to $\times 200$ magnification. The assay can be scored at any time after syncytia are readily apparent in untreated control wells, with syncytia being defined as four or more nuclei within a clear common cell membrane. At relatively early time points following mixing of cell populations, varying degrees of cell fusion may be seen and can be scored on a 0 to 4+ scale, as described elsewhere.[24,25] After 24 hr of cocultivation syncytia formation has generally proceeded to its full extent, and, if any syncytia are present in the well, cell fusion is typically extensive enough to warrant a 4+ score. Thus, at 24 hr after cocultivation, wells may generally be scored as positive or negative, without a need for intermediate scores. Even in the presence of serial dilutions of inhibitors, fusion usually appears to be an all-or-nothing phenomenon. When the assay is scored at 24 hr after cocultivation, it is typical to observe an abrupt transition from seeing no syncytia with a $1:2^n$ dilution of inhibitor to seeing extensive (4+) syncytia with a $1:2^{n+1}$ dilution of the same inhibitor. This feature allows definition of an end-point dilution for complete inhibition of cell fusion for each inhibitor. This parameter tends to be highly reproducible; for given inhibitors tested in multiple separate assays, even assays months apart, the end-point dilution is generally within a factor of 2–4.

Other Scoring Methods. In our hands, simple scoring by inverted phase-contrast microscopy for the presence or absence of unmistakable

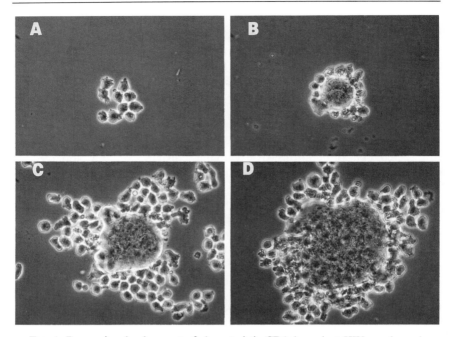

FIG. 1. Progressive development of characteristic CD4-dependent HIV envelope glyco-protein-induced multinucleated giant cells on cocultivation of VB cells with H9 cells chroni-cally infected with HIV-1$_{HXB-2}$ viewed by phase-contrast microscopy. (A) Appearance of a cluster of cells immediately after mixing. Cells are indistinguishable from uninfected cells. (B) Appearance of a relatively early multinucleated giant cell. The central cell mass has incorpo-rated 5–10 nuclei via cell fusion and is surrounded by a rosette of attached cells in the process of fusing to the central cell mass. (C) Well-developed multinucleated giant cell. (D) Large multinucleated giant cell that has incorporated upward of 100 nuclei and is still surrounded by additional attached cells in the process of fusing. By 24 hr of culture, this cell will have undergone extensive degenerative changes, with nuclear pyknosis and ballooning degenera-tion of the cytoplasm. (From Lifson and Engleman[22] with permission. © 1989 Munksgaard International Publishers Ltd., Copenhagen, Denmark.)

characteristic HIV-induced syncytia (Fig. 1) is an accurate, highly repro-ducible method for evaluating CD4-dependent, HIV envelope-induced binding/cell fusion events, including the quantitative testing of inhibitors of these processes. Scoring by microscope is straight forward, rapid, and allows morphological examination of the cells, permitting assessment of any cytotoxic effects attributable to putative inhibitors undergoing evalua-tion. Addition of trypan blue to wells prior to microscopic evaluation may increase the sensitivity of microscopic examination as an approach for assessment of cytotoxic effects. In addition, because of its simplicity and convenience, this approach allows relatively high throughput testing of compounds, for a nonautomated assay, with up to hundreds of compounds evaluable per assay by a single individual over a 24-hr period. Although syncytia present in each test well may be enumerated, we have not found this procedure to be either necessary or particularly rewarding, especially given the reliability and reproducibility of end-point titers as a means of quantitating the potency of inhibitors. Similarly, we have not found it necessary to employ additional instrumentation to quantitate the number or size of HIV-induced syncytia.

Approaches to HIV-Induced Cell Fusion Not Requiring Infectious HIV

As noted above, the assay procedure described above requires the use of HIV-infected cells, with all the attendant biosafety implications.[37,38] Although some investigators may not have access to facilities necessary for handling infectious HIV, this does not absolutely preclude performing assays similar to that described above. Work from numerous investigators has demonstrated that cell surface expression of HIV envelope glycoproteins (gp120 and gp41) and CD4 is necessary and sufficient to mediate characteristic HIV-induced cell fusion.[24-26] Thus expression of the HIV envelope glycoprotein gene apart from other HIV genes can be used as an approach toward studying these interactions without the requirement of handling infectious HIV. This has been achieved in recombinant vaccinia virus systems,[44,45] which have been usefully employed in the analysis of HIV-induced cell fusion.[25] It should be noted, however, that handling of vaccinia viruses has its own biosafety implications and personnel involved in such studies should have appropriate training and vaccination histories. Because vaccinia virus infection is itself cytopathic for susceptible cells, it is not possible, using this approach, to generate cell populations analogous to the chronically HIV-infected H9 cells used in the assay procedure described above. However, other recombinant DNA approaches can be employed to generate cells constitutively expressing HIV envelope glycoproteins in the absence of other HIV genes. These cells can be readily used to replace the HIV-infected H9 cells used in the assay procedure described above.

Summary

CD4-dependent HIV envelope glycoprotein-induced membrane fusion events play a key role in the life cycle of HIV and are involved both in

---

[44] S. Chakrabarti, M. Robert-Guroff, F. Wong-Staal, R. C. Gallo, and B. Moss, *Nature (London)* **320,** 535 (1986).
[45] P. Ashorn, E. A. Berger, and B. Moss, [2] this volume.

infection mediated by viral particles and in virally mediated cytopathic processes. The relevant events involve binding interactions between the HIV envelope glycoprotein gp120 and the cellular receptor CD4 and membrane fusion processes mediated by the HIV envelope glycoprotein gp41. A straight forward, rapid, and convenient assay procedure useful for analysis of these processes and identification of inhibitors is described.

Acknowledgments

This work was supported in part by NIH Grant AI/CA-25922. The author thanks E. G. Engleman, L. E. Eiden, K. M. Hwang, and A. E. Keylor for their contributions to the work described here, and G. R. Reyes for thoughtful review of the manuscript.

[2] Vaccinia Virus Vectors for Study of Membrane Fusion Mediated by Human Immunodeficiency Virus Envelope Glycoprotein and CD4

By PER ASHORN, EDWARD A. BERGER, and BERNARD MOSS

Introduction

Vaccinia virus has been widely used to express genes from a variety of sources.[1] In general, synthesis, transport, processing, and modification of proteins occur normally in mammalian cells infected with recombinant vaccinia viruses. The surface expression of viral envelope proteins by vaccinia virus vectors has led to their use for the study of the fusion activity of human immunodeficiency virus (HIV) type 1[2] and type 2,[3] simian immunodeficiency virus,[4] as well as bovine parainfluenza virus[5] and vesicular stomatitis virus.[6]

[1] B. Moss and C. Flexner, *Annu. Rev. Immunol.* **5,** 305 (1987).

[2] J. Lifson, M. Feinberg, G. Reyes, L. Rabin, B. Banapour, S. Chakrabarti, B. Moss, F. Wong-Staal, K. Steimer, and E. Engelman, *Nature (London)* **323,** 725 (1986).

[3] S. Chakrabarti, T. Mizukami, G. Franchini, and B. Moss, *Virology* **178,** 124 (1990); M. J. Mulligan, P. Kumar, H. Hui, R. J. Owens, G. D. Ritter, B. H. Hahn, R. W. Compans, *AIDS Res. Hum. Retroviruses* **6,** 707 (1990).

[4] M. L. Bosch, P. L. Earl, K. Fargnoli, S. Picciafuoco, F. Giombini, F. Wong-Staal, and G. Franchini, *Science* **244,** 694 (1989).

[5] Y. Sakai and H. Shibuta, *J. Virol.* **63,** 3661 (1989).

[6] M. Whitt, P. Zagouras, B. Cruse, and J. K. Rose, *J. Virol.* **64,** 4907 (1990).

Recombinant vaccinia virus technology has been used to study membrane fusion mediated by HIV-1 envelope glycoprotein (Env) and its cellular receptor CD4. Human immunodeficiency virus infects human cells by binding to their surface CD4 molecules and directly fusing with the cell membrane.[7-9] In contrast, murine cells expressing human CD4 bind to HIV, but do not become infected.[7] This block to infection seems to occur in the virus internalization process, because murine cells have been shown to support HIV replication after transfection with a genomic provirus.[10] Syncytium formation (cell–cell fusion) is used as a model for virus–cell fusion, and utilizes cocultures of human and nonhuman cells expressing the appropriate surface molecules encoded by vaccinia virus vectors. The formation of multinucleated giant cells was initially observed in HIV-infected helper T lymphocytes[11] and has been subsequently demonstrated in cultures of CD4-bearing human lymphocytes expressing Env gp and no other HIV proteins.[2] A variety of human and nonhuman cell lines can be rapidly screened for their fusion capabilities, using recombinant vaccinia viruses to express high levels of CD4 or HIV-1 Env transiently and then assaying syncytium formation between cells bearing these surface proteins.

Cell Culture and Recombinant Vaccinia Viruses

Suspension cell medium contained RPMI 1640 (Quality Biologicals, Rockville, MD) supplemented with 10% (v/v) fetal calf serum (FCS; GIBCO, Grand Island, NY), 2 mM L-glutamine, 10 mM N-2-hydroxyethylpiperazine-N'-2-ethanesulfonic acid (HEPES), and antibiotics. Growth medium for adherent cell lines containing Dulbecco's modified Eagle's medium (DMEM; Quality Biologicals) is supplemented with 10% (v/v) FCS, 2 mM glutamine, and antibiotics. Cells are maintained at 37° in a 5% CO_2 atmosphere.

General procedures for the construction of recombinant vaccinia viruses have been described in detail.[12,13] A recombinant vaccinia virus that directs the expression of the full-length human CD4 molecule under the

[7] P. Maddon, A. Dalgleish, J. McDougal, P. Clapham, R. Weiss, and R. Axel, *Cell (Cambridge, Mass.)* **47**, 333 (1986).

[8] B. Stein, S. Gowda, J. Lifson, R. Penhallow, K. Bensch, and E. Engleman, *Cell (Cambridge, Mass.)* **49**, 659 (1987).

[9] M. McClure, M. Marsh, and R. Weiss, *EMBO J.* **7**, 513 (1988).

[10] J. Levy, C. Cheng-Mayer, D. Dina, and P. Luciw, *Science* **232**, 998 (1986).

[11] M. Popovič, M. Sarngadharan, E. Read, and R. C. Gallo, *Science* **224**, 497 (1984).

[12] M. Mackett, G. L. Smith, and B. Moss, *in* "DNA Cloning" (D. M. Glover, ed.), Vol. 2, p. 191. IRL Press, Oxford, 1985.

[13] P. Earl and B. Moss, *in* "Current Protocols in Molecular Biology" (F. M. Ausubel *et al.,* eds.), Vol. 2, suppl. 15, pp. 16.17.1–16.17.16. Wiley, New York.

control of a bacteriophage T7 promoter is constructed.[14] Expression of CD4 is achieved on coinfection of cells with the latter virus, as well as a second recombinant vaccinia virus that contains the bacteriophage T7 RNA polymerase gene under control of a vaccinia virus promoter.[15] Another recombinant virus, with a vaccinia promoter regulating the full-length HIV-1 *env* gene, induces the synthesis of gp160, which is correctly processed to gp120 and gp41 subunits.[16] As a control, a recombinant virus with *Escherichia coli* β-galactosidase under control of a vaccinia promoter is used.[17]

Transient Expression of CD4 and HIV-1 Envelope Glycoprotein

Suspension cells are washed once with medium and suspended at 10^7 cells/ml; then recombinant vaccinia viruses are added at a multiplicity of 10 plaque-forming units (pfu)/cell. After a 1-hr adsorption period, cells are diluted to a density of 5×10^5 cells/ml and placed in a CO_2 incubator for 10 to 14 hr. Adherent cells are trypsinized, washed twice with suspension cell medium, and infected in suspension. Surface expression of CD4 and HIV-1 envelope glycoprotein is monitored by immunofluorescence microscopy and with a fluoresecence-activated cell sorter (FACS).

Syncytium Formation Assay

Cells infected with recombinant vaccinia viruses and incubated for 10 to 14 hr are washed twice with phosphate-buffered saline (PBS) and suspended at 10^6 cells/ml in suspension cell medium. One-half milliliter of HIV-1 Env-expressing cells is then mixed with 0.5 ml of CD4-bearing cells in 24-well tissue culture plates (Costar, Cambridge, MA). In control experiments, cells infected with vaccinia virus expressing *E. coli* β-galactosidase are substituted for CD4- or Env-expressing cells. For fusion inhibition studies, cells expressing CD4 or HIV-1 Env are preincubated for 1 hr with sCD4 (soluble CD4, comprising the four extracellular domains of the protein) [obtained from S. Johnson (The UpJohn Company, Kalamazoo, MI)], or monoclonal antibody OKT4A (Ortho Diagnostics, Raritan, NJ), before adding the fusion partner to the culture. In coculture studies, syncytia appear at 1 to 3 hr after mixing of the appropriate fusion partners.

---

[14] P. Ashorn, E. A. Berger, and B. Moss, *J. Virol* **64**, 2149 (1990).
[15] T. Fuerst, P. Earl, and B. Moss, *Mol. Cell. Biol.* **7**, 2538 (1987).
[16] P. L. Earl, A. Hügin, and B. Moss, *J. Virol.* **64**, 2448 (1990).
[17] S. Chakrabarti, K. Brechling, and B. Moss, *Mol. Cell. Biol.* **5**, 3403 (1985).

HIV Envelope Glycoprotein-Mediated Formation of Human Cell Syncytia

The model is tested initially by monitoring the syncytium formation in CD4-negative A2.01 human T cells[18] infected with recombinant vaccinia viruses that express CD4 and HIV-1 Env. When infected A2.01 cells expressing Env are mixed with infected A2.01 cells expressing CD4, large syncytia appear in culture within 1 to 3 hr. The specificity of the reaction is indicated by the ability of sCD4, or monoclonal antibody OKT4A, to inhibit syncytium formation, and by the inability of cells infected with a control vaccinia virus to fuse with Env- or CD4-bearing cells.

HIV-1 Envelope Glycoprotein-Mediated Formation of Human – Nonhuman Cell Syncytia

A variety of lymphoid and nonlymphoid animal cell lines (listed in Table I) are infected with recombinant vaccinia viruses encoding CD4 or Env. As with human cells, significant protein expression on the cell surface is observed by immunofluorescence at 10 to 14 hr after the infection of all cell lines tested.

The species specificity of HIV Env-mediated membrane fusion is explored by conducting mixing experiments with Env-expressing and CD4-bearing animal and human cells. As shown in Table I, all the tested cell lines expressing Env form syncytia when mixed with CD4-positive human cells. In each case syncytium formation is significant, although less efficient than when mixing human cells only. As with human – human syncytia, specificity is indicated by the inhibition of syncytium formation by soluble CD4 or OKT4A antibody. With three exceptions, cell fusion is unidirectional, that is, it occurs only if CD4 is expressed on human cells (Table I). No syncytia are observed if both Env and CD4 are expressed on any of the nonhuman cells.

Conclusions

Because the internalization of HIV occurs via direct fusion of viral and cell membranes, syncytium formation is probably mediated through a similar process.[2,7,8] A rapid screen of a large number of human and nonhuman cells for their fusion characteristics can be made by using recombinant vaccinia viruses, encoding either human CD4 or HIV Env. This

[18] T. Folks, D. M. Powell, M. M. Lightfoote, S. Benn, M. A. Martin, and A. S. Fauci, *Science* **231**, 600 (1986).

TABLE I

SYNCYTIUM FORMATION BETWEEN HUMAN AND NONHUMAN CELLS
EXPRESSING CD4 AND HIV-1 ENVELOPE GLYCOPROTEIN[a]

Cell line			Syncytium formation[b]		
Name	Species	Type	Cell line/ Env + A2.01/CD4	Cell line/ CD4 + A2.01/Env	Cell line/ Env + Cell line/CD4
A2.01	Human	T lymphocyte	+	+	+
TK143⁻	Human	Fibroblast	+	+	+
CV-1	AGM[c]	Kidney	+	−	−
26 CB-1	Baboon	Lymphoblast	+	−	−
6056	Rabbit	Macrophage	+	−	−
6516	Rabbit	T lymphocyte	+	−	−
RL-5	Rabbit	T lymphocyte	+	+	−
RK13	Rabbit	Kidney	+	−	−
E 36	Hamster	Lung	+	−	−
Y3-Ag1.2.3	Rat	Plasma cell	+	+	−
BW 5147	Mouse	T lymphocyte	+	−	−
A-20	Mouse	B lymphocyte	+	−	−
SP-2	Mouse	Plasma cell	+	−	−
PU 5-1.8	Mouse	Macrophage	+	+	−
P.815	Mouse	Mast cell	+	−	−
NIH 3T3	Mouse	Fibroblast	+	−	−
MC 57	Mouse	Fibroblast	+	−	−
CL-7	Mouse	Fibroblast	+	−	−
L-A9	Mouse	Fibroblast	+	−	−
STO	Mouse	Fibroblast	+	−	−
TCMK-1	Mouse	Kidney	+	−	−
CMT-93	Mouse	Rectum carcinoma	+	−	−

[a] Encoded by recombinant vaccinia virus vectors.

[b] Giant cell formation at 3 hr after mixing HIV-1 Env-expressing cells and CD4-bearing human lymphocytes (cell line/Env + A2.01/CD4), CD4-bearing cells and Env-expressing human lymphocytes (cell line/CD4 + A2.01/Env) or Env-expressing and CD4-bearing cells (cell line/Env + cell line/CD4) in tissue culture. A positive (+) result indicates that the culture contained cells with a diameter three times larger than that of a single cell. A negative (−) result indicates that the culture contained only single cells.[14]

[c] AGM, African green monkey.

technique has revealed the following: (1) CD4-bearing lymphoid and non-lymphoid human cells readily fuse to human cells expressing HIV-1 Env; (2) CD4-bearing nonhuman cells invariably fail to fuse with Env-expressing cells of the same type; (3) Env-bearing nonhuman cells always fuse with human cells expressing CD4; and (4) in only 3 of 20 cases, syncytium formation occurred when CD4 was expressed on the nonhuman cell and the Env-expressing cell was of human origin.

The data provide some insights into the failure of murine and other nonhuman cells to internalize HIV.[14] It is unlikely that the internalization block is due to any inherent interspecies membrane incompatibility, because (1) HIV-1 Env-expressing nonhuman cells do fuse with CD4-bearing human cells and (2) syncytia do not form in cultures containing CD4-bearing and Env-expressing cells of the same nonhuman cell type, a situation in which membrane incompatability does not exist.

A requirement for a second human cell membrane protein that interacts either (trans) with HIV-1 Env or (cis) with the CD4 receptor may be another reason for the inability of nonhuman cells to internalize HIV. There are, however, three exceptional rabbit, rat, and mouse cell lines that can fuse with a human cell, regardless of which partner expresses CD4. In addition, mouse–human T cell hybrids, containing all human chromosomes and properly expressing CD4, are not infectable by HIV.[19]

A third explanation for the unidirectionality of nonhuman cell–human cell fusion is related to differences in lipid composition of cell membranes. According to this model, the fusion peptide at the NH_2-terminus of gp41 can interact only with human membranes. This model, however, does not account for the three exceptions to unidirectionality. The CD4–Env fusion problem may be a multifaceted process in which successful fusion may depend on several factors, whose influences vary among different cell types. Thus, in certain cases, fusion may be limited not by the mere presence of CD4 and Env, but by other interrelated membrane properties, such as auxiliary adhesion components, surface charge, membrane fluidity, and the mobility of specific surface molecules within the membrane.

Comments

When separate batches of cells are infected with different recombinant vaccinia viruses, it is necessary to rule out the possibility of virus spread after mixing the cells. Experiments must be done to demonstrate that the syncytia contain both human and nonhuman cells. Labeling the nonhuman cells with a fluorescent dye, fluorescein isothiocyanate (FITC), demonstrates that the syncytia contain not only human, but also nonhuman, cells. Virus spread appears to be unlikely, because an exclusion phenomenon largely prevents the superinfection of cells already infected with vac-

[19] M. Tersmette, J. van Dongen, P. Clapham, R. de Goede, I. Wolvers-Tettero, A. Geurts van Kessel, J. Huisman, R. Weiss, and F. Miedema, *Virology* **168,** 267 (1989).

cinia virus. In the assay conditions described here, syncytium formation occurs rapidly, before significant amounts of protein can be expressed from superinfected cells.

Under low-pH conditions, syncytium formation is observed when cells are infected with some strains of vaccinia virus. This can be detected at pH 6.4 and may be extensive at pH 5.8.[20,21] When studying viruses such as HIV, whose entry into the host cells is pH independent, this fusogenic characteristic of vaccinia virus is not a significant problem. Such low pH values are inconsistent with prolonged cell viability and are unlikely to occur spontaneously. pH drops can be avoided by adding HEPES buffer to the medium and maintaining the cells at a fairly low density (not to exceed 0.5×10^6/ml).

The vaccinia-based expression system provides a powerful tool for further investigating the membrane fusion mediated by HIV, or other viral envelope glycoproteins and their receptors. It allows the expression of both of these molecules, in conjunction with other relevant surface proteins, in a broad range of cell types. This provides an opportunity to test the effects of expression of auxiliary adhesion components as well as mutations in CD4 and HIV-1 Env that may influence membrane mobility and the capacity of receptor–ligand interaction to promote membrane fusion.

[20] K. Kohono, J. Sambrook, and M.-J.J. Gething, *J. Cell. Biochem., Suppl.* **12,** 29 (1988).
[21] R. W. Doms, R. Blumenthal, and B. Moss, *J. Virol.* **64,** 4884 (1990).

[3] Sendai Virus-Induced Cell Fusion

By YOSHIO OKADA

Introduction

Sendai virus [also referred to as hemagglutinating virus of Japan (HVJ)] was the first agent found to be effective in the fusion of cells. The cell fusion activity of HVJ particles was first reported in 1957[1] and has since been used to study cell biology and cell genetics. The cell fusion activity of HVJ is resistant to ultraviolet (UV) irradiation and can be induced by UV-inactivated HVJ that cannot grow.[2]

The discovery of HVJ-induced cell fusion coincided with the establishment of methods for *in vitro* culture of somatic cells derived from mammals and fowls. Since 1965, cell fusion has been used extensively in studies in cell biology and the field of "somatic cell genetics" has been established.

[1] Y. Okada, Y. Hosaka, and T. Suzuki, *Med. J. Osaka Univ.* **7,** 709 (1957).
[2] Y. Okada and J. Tadokoro, *Exp. Cell Res.* **26,** 108 (1962).

Some examples of progress in this field include heterokaryon formation,[3,4] somatic cell hybrid formation from heterokaryons,[5] the establishment of an effective system for selection of hybrid cells by the use of tk⁻ and hprt⁻ mutants,[6] reactivation of dormant nuclei derived from chick erythrocytes in heterokaryons fused with cultured mammalian cells,[7,8] utilization of cell fusion for cell differentiation analysis,[9,10] induction of transcription of tumor virus genomes integrated into transformed nonproducer cells by their fusion with uninfected permissive host cells,[11,12] the demonstration of instability of chromosomal balance in interspecific hybrid cells[13] and of selective disappearance of human chromosomes in human–mouse hybrid cells in culture[14] (which led to the field of gene mapping on human chromosomes),[15] and detection of complementary groups in a hereditary disease, xeroderma pigmentosum.[16]

Polyethylene glycol (PEG), an effective fusogenic agent, was discovered by Kao and Michayluk in 1974,[17] and in the 1980s the electroporation technique was found to be useful for cell fusion.[17a] With the establishment of these two methods, cell fusion became applicable to the cells of invertebrates and plants that have no HVJ receptors. Polyethylene glycol has been especially useful for fusion of B lymphocytes with myeloma cells and for the preparation of cultures producing monoclonal antibodies.[18] Since 1974, certain refinements of the cell fusion reaction have been introduced, and cells have been reconstituted by fusion of nucleoplasts and cytoplasts separated by treatment with cytochalasin.[19] This procedure has been used

[3] H. Harris and J. F. Watkins, *Nature (London)* **205**, 640 (1965).

[4] Y. Okada and F. Murayama, *Biken J.* **8**, 7 (1965).

[5] G. Yerganian and M. B. Nell, *Proc. Natl. Acad. Sci. U.S.A.* **55**, 1066 (1966).

[6] J. Littlefield, *Exp. Cell Res.* **41**, 190 (1966).

[7] H. Harris, *Nature (London)* **206**, 583 (1965).

[8] H. Harris, E. Sidebottom, D. M. Grace, and M. E. Bramwell, *J. Cell Sci.* **4**, 499 (1969).

[9] B. W. Finch and B. Ephrussi, *Proc. Natl. Acad. Sci. U.S.A.* **57**, 615 (1967).

[10] R. L. Davidson and K. Yamamoto, *Proc. Natl. Acad. Sci. U.S.A.* **60**, 894 (1968).

[11] P. Gerber, *Virology* **28**, 501 (1966).

[12] J. Svoboda, O. Machala, and I. Hlozanek, *Acta Virol.* **13**, 155 (1967).

[13] M. C. Weiss and B. Ephrussi, *Genetics* **54**, 1095 (1966).

[14] M. C. Weiss and H. Green, *Proc. Natl. Acad. Sci. U.S.A.* **58**, 1104 (1967).

[15] F. H. Ruddle and R. P. Creagan, *Ann. Rev. Genet.* **9**, 407 (1975).

[16] E. A. deWeerd-Kastelein, W. Keijzer, and D. Bootsma, *Nature (London), New Biol.* **238**, 80 (1972).

[17] K. N. Kao and M. R. Michayluk, *Planta* **115**, 355 (1974).

[17a] M. Senda, J. Takeda, S. Abe, and T. Nakamura, *Plant Cell Physiol.* **20**, 1441–1443 (1979).

[18] G. Köhler and C. Milstein, *Nature (London)* **256**, 495 (1975).

[19] G. Veomett, D. M. Prescott, J. Shay, and K. R. Porter, *Proc. Natl. Acad. Sci. U.S.A.* **71**, 1999 (1974).

to introduce nuclei into enucleated eggs to obtain genetically altered animals.[20] Cell fusion also allows the introduction of macromolecules into living cells.[21-23]

In this chapter, the mechanism and details of cell fusion by HVJ are described.

Structure and Characteristics of Sendai Virus

Sendai virus (HVJ), a paramyxovirus, is the causal agent of murine pneumonitis. The virus is an enveloped virus that contains six kinds of proteins (Table I) and a negative strand of RNA. The F and HANA (HN) glycoproteins are exposed as spikes on the envelope. The M protein connects an internal nucleocapsid with the envelope. This nucleocapsid consists of nucleocapsid proteins (NP), genomic RNA, and the L and P proteins, and has a left-handed helical structure of about 1 μm in length and 17 to 18 nm in width. The L and P proteins are believed to have RNA-dependent RNA polymerase activity. The C protein is not an integral part of the virus, but is expressed in infected host cells. Its function is not yet clear.

The total length of the genomic RNA is 15,383 nucleotides. The 3' terminal and 5' terminal are the leader regions, in which there are six open reading frame shifts for NP, P(C), M, F, HN, and L, in that order. The sequence of the C gene is encoded in the P gene and is transcribed by frame shift. Each reading frame is flanked by a consensus sequence R_1 at the 3' end and R_2 at the 5' end.[24]

Mode of Infection and Fusion Activity of Sendai Virus

Sendai virus (HVJ) is the first animal virus whose mode of infection has been clarified. Infection occurs by infusion of the viral envelope with the cell membrane, and the introduction of the nucleocapsid into the cytoplasm of the host cells.[25] The fusion step is essential for infection and the fusion activity is closely related to the hemolytic and cell-to-cell fusion activities. All three activities are lost when the envelopes are solubilized with detergent, and reappear when the structure is reassembled on removal of the detergent.[26] The molar ratio of active F protein to HANA protein

[20] J. McGrath and D. Solter, *Science* **220**, 1300 (1983).
[21] M. Furusawa, T. Nishimura, M. Yamaizumi, and Y. Okada, *Nature (London)* **249**, 449 (1974).
[22] K. Tanaka, M. Sekiguchi, and Y. Okada, *Proc. Natl. Acad. Sci. U.S.A.* **72**, 4071 (1975).
[23] T. Uchida, J. Kim, M. Yamaizumi, Y. Miyake, and Y. Okada, *J. Cell Biol.* **80**, 10 (1979).
[24] T. Shioda, K. Iwasaki, and H. Shibuta, *Nucleic Acids Res.* **14**, 1545 (1986).
[25] C. Morgan and C. J. Howe, *J. Virol.* **2**, 1122 (1968).
[26] Y. Hosaka and Y. K. Shimizu, *Virology* **49**, 627 (1972).

TABLE I
PROTEINS OF SENDAI VIRUS

Designation	Molecular weight[a]	Function
L	224	RNA polymerase
P	79	RNA polymerase
HANA (HN)[b]	72 (65–74)[c]	Hemagglutinin, neuraminidase
F_0^b	64.7 (65–68)[c]	Fusion (inactive form)
F_1^b	51.5 (47–56)[c]	Fusion (active form)
F_2^b	11.3 (10–16)[c]	Fusion (active form)
NP	60	Nucleocapsid protein
M	34	Connection of nucleocapsid with envelope
C	22	Nonstructural protein

[a] Molecular weight (kDa) deduced from cDNA.
[b] Containing carbohydrates.
[c] Molecular weight given in parentheses is deduced from mobility on electrophoresis.

integrated into reassembled envelopes greatly affects the efficiency of the fusion reaction with cell membranes; the most effective molar ratio is F : HANNA = 2 : 1, which is similar to their ratio in virions.[27]

Growth of Virus Stock in Embryonated Chicken Eggs

As virus stock for cell fusion, HVJ is harvested from the chorioallantoic fluid (CAF) of 10-day-old embryonated chicken eggs that have been incubated for 3 days at 38° after injection of seed virus (0.1 ml of 10^3-fold diluted infected CAF). The yield of this virus is high, being on the order of 10^{11} virus particles per egg. The virus particles propagated in embryonated eggs are fully infective, having hemagglutinating, hemolytic, and cell fusion activities.[28] The titer of the virus is determined by Salk's pattern method and expressed as hemagglutinating units (HAU). One hemagglutinating unit corresponds to about 2.4×10^7 virus particles.[29] On long-term passages viral progeny with low cell fusion activity may appear, and if so the seed virus must be replaced by old stock.[30]

Characteristics of Sendai Virus Grown in Cultured Cells

On infection of monolayer cultures of susceptible cells, the virus does not form plaques but shows only one step of growth; the virus progeny

[27] M. Nakanishi, T. Uchida, and Y. Okada, *Exp. Cell Res.* **142,** 95 (1982).
[28] K. Fukai and T. Suzuki, *Med. J. Osaka Univ.* **6,** 1 (1955).
[29] Y. Okada, S. Nishida, and J. Tadokoro, *Biken J.* **4,** 209 (1961).
[30] Y. Okada and Y. Hosokawa, *Biken J.* **4,** 217 (1961).

have hemagglutinating activity and are adsorbed to the surface of host cells, but have no ability to infect cultured cells.[31,32] They also show no cell fusion or hemolytic (HL) activity. However, these progeny are not incomplete, because they appear fully capable of infecting embryonated eggs. These progeny cannot infect cultured cells because they have inactive F glycoprotein (F_0), and the cleavage of F_0 to F_1 and F_2 by some proteolytic enzyme, such as trypsin, is essential for infectivity.[33,34] When injected into the chorioallantoic cavity of embryonated eggs, the cleavage of F_0 proceeds in the absence of an added proteolytic enzyme.[35] The proteolytic enzyme present in CAF was identified as factor X of the blood coagulation system.[36] With the cleavage of F_0, HVJ acquires cell fusion activity, HL activity, and infectivity, all simultaneously.

F Glycoprotein and Its Function

Figure 1 shows the base sequence of F_0 cDNA and the hydrophobic regions of the peptide deduced from the base sequence.[37] The peptide has three hydrophobic regions. From the N terminus, the first is a signal sequence (amino acids 11 to 23), the second is a fusogenic domain, and the third (amino acids 500 to 523) is a membrane anchorage domain. The C-terminal region (42 amino acids) is the viroplasmic domain, which may connect with the M protein. There are three potential glycosylation sites. F_0 is activated by proteolytic cleavage, between Arg-116 and Phe-117, into two parts, F_2 and F_1. By this cleavage, the hydrophobic fusogenic domain (amino acids 117 to 142) is exposed at the new N terminus of F_1. The sequence of the fusogenic domain is well conserved in paramyxoviruses. The sequence Pro-Gln-Ser-Arg (amino acids 113 to 116) upstream of the fusogenic domain is believed to be the cleavage recognition site. This sequence differs in various paramyxoviruses and is related to their virulence. The sequence of HVJ is typical for an avirulent virus.[38] The fusogenic domain has the capacity to trap cholesterol at 37°, but not at 20°.[39]

[31] N. Ishida and M. Homma, *Virology* **14**, 486 (1961).
[32] T. Matsumoto and K. Maeno, *J. Virol.* **8**, 722 (1962).
[33] M. Homma, *J. Virol.* **8**, 619 (1971).
[34] M. Homma, *J. Virol.* **9**, 829 (1972).
[35] M. Homma and M. Ohuchi, *J. Virol.* **12**, 1457 (1973).
[36] B. Goto, T. Toyoda, M. Hamaguchi, and Y. Nagai, *Abstr. 37th Meet. Jpn. Virol.*, p. 162 (1989).
[37] N. Miura, E. Ohtsuka, N. Yamaberi, M. Ikehara, T. Uchida, and Y. Okada, *Gene* **38**, 271 (1985).
[38] Y. Okada, Y. Shima, T. Shimamoto, N. Kusaka, and Y. Kiho, *Cell Struct. Funct.* **14**, 707 (1989).
[39] K. Asano and A. Asano, *Biochemistry* **27**, 1321 (1988).

HANA Glycoprotein

HANA is a glycoprotein with hemagglutination and neuraminidase activities. The cellular virus receptor consists of sialoglycoproteins and sialolipids, and HANA contains both receptor-binding and receptor-degrading activities. The minimal structure recognized by HANA is the sequence NeuAcα2,3Gal.[40] The peptide deduced from the cDNA of HANA consists of 575 amino acids and its N terminus (amino acids 35 to 60), which is the only highly hydrophobic region, may be both the signal sequence and the membrane anchorage domain.[41] Five potential acceptor sites for N-linked carbohydrates are detectable in this peptide sequence.

Hemolytic Activity and Aging of Virions

Virions in stocks harvested from embryonated eggs show the characteristic of pleomorphism, and vary in both size and shape. The smallest virions have one nucleocapsid, whereas larger ones have various numbers of nucleocapsids, depending on their size.[42] When envelopes of virions are freeze-fractured, large intramembrane particles (IMPs) of about 150 Å in diameter can be seen on the outer leaflet (E fracture face) of large virions, but no IMPs can be seen on either the inner leaflet (P fracture face) or E fracture face of small virions. Judging from their molecular size, the transmembrane domains of both F and HANA glycoproteins should be invisible. Immediately after budding, nonaged virions resemble the smallest virions observed in the virus stock, being small in size and having no detectable IMPs on either fracture face. On incubation at 37° for 2 days *in vitro,* large IMPs become visible in the aged virions. These findings suggest that large IMPs are formed during aging, by aggregation of the transmembrane domains of the glycoproteins.[43]

This change seems to be induced by dissociation of the viroplasmic domain of the glycoprotein molecules and M protein, resulting in aggregation of their transmembrane domains, and the movement of the envelope from the P fracture face to the E fracture face of the lipid bilayer. The aggregation of the membrane domains should induce naked areas in the viral envelope, and spontaneous fusion of aged virions by attachment of the naked areas to those of adjacent virions. In fact, fusion of aged virions is observed when concentrated virions are incubated at 37°.[44]

[40] M. A. K. Markwell, L. Svennerholm, and J. C. Paulson, *Proc. Natl. Acad. Sci. U.S.A.* **78,** 5406 (1981).
[41] N. Miura, Y. Nakatani, M. Ishiura, T. Uchida, and Y. Okada, *FEBS Lett.* **188,** 112 (1985).
[42] Y. Hosaka and H. Kitano, *Virology* **29,** 205 (1966).
[43] J. Kim, K. Hama, Y. Miyake, and Y. Okada, *Virology* **95,** 523 (1979).
[44] J. Kim and Y. Okada, *J. Membr. Biol.* **97,** 241 (1987).

```
 936-  GTC CCA GGT GTG CTC ATA CAC AAG GCA TCA TCT ATT TCT TAC AAC ATA GAC GGG GAG GAA TGG TAT GTG ACT GTC CCC AGC CAT  -322
       Val Pro Gly Val Leu Ile His Lys Ala Ser Ser Ile Ser Tyr Asn Ile Asp Gly Glu Glu Trp Tyr Val Thr Val Pro Ser His

1020-  ATA CTC AGT CGT TTC TTA GGG GGT GCA ATA ACC GAT TGT GTT GAG TCC AGA TTG ACC TAT ATA TGC CCC AGG GAT  -350
       Ile Leu Ser Arg Phe Leu Gly Gly Ala Ile Thr Asp Cys○ Val Glu Ser Arg Leu Thr Tyr Ile Cys Pro Arg Asp

1104-  CCC GCA CAA CTG ATA CCT GAC AGC AGC CAG CAA AAG TGT ATC CTG GGG ACA ACA AGG TGT CCT GTC ACA AAA GTT GTG GAC AGC  -378
       Pro Ala Gln Leu Ile Pro Asp Ser Ser Gln Gln Lys Cys Ile Leu Gly Thr Thr Arg Cys○ Pro Val Thr Lys Val Val Asp Ser

1188-  CTT ATC CCC AAG TTT GCT TTT GTG AAT GGG GTT GGC GTT GTA GCA TCC ATA TGC ACA TGT ACC GGG ACA GGC CGA AGA  -406
       Leu Ile Pro Lys Phe Ala Phe Val Asn Gly Val Gly Val Val Ala Ser Ile Cys Thr Cys○ Thr Gly Thr Gly Arg Arg

1272-  CCA ATC AGT CAG GAT CGC TCT AAA GGT GTA TTC CTA ACC CAT GAC AAC TGT GGT CTT ATA GGT GTC AAT GGG GTA GAA TTG  -434
       Pro Ile Ser Gln Asp Arg Ser Lys Gly Val Phe Leu Thr His Asp Asn Cys○ Gly Leu Ile Gly Val Asn Gly Val Glu Leu

1356-  TAT GCT AAC CGG AGA GGG CAC GAT GCC ACT TGG GTC CAG GTC ACA TTG AAC TTG ACA GTC GGT CCT GCA ATT GCT ATC AGA CCC ATT GAT  -462
       Tyr Ala Asn Arg Arg Gly His Asp Ala Thr Trp Val Gln Val Thr Leu | Asn Leu Thr | Val Gly Pro Ala Ile Ala Ile Arg Pro Ile Asp

1440-  ATT TCT CTC AAC CTT GCT GAT GCT ACG AAT TTC TTG CAA GAC CTT TGG CTT CAA GAC CTT AAG GTA GTT ATG GTC ATG GAC GTA GAC AGA CGG GCA AAA  -490
       Ile Ser Leu Asn Leu Ala Asp Ala Thr Asn Phe Leu Gln Asp Leu Trp Leu Gln Asp Leu Lys ... Glu Lys Ala Arg Lys

1524-  GTA GGT TAC TGG TAC AAC TCA AGA ACT GTG AGG GAG ATC ACG GTA GTT ATT ACG GTC ATA TTG GTC GTC ATA ATA GTG ATC  -518
       Val Gly Tyr Trp Tyr Asn Ser Arg Thr Val Arg Glu Ile Thr Val Val Ile Thr Val Ile Leu Val Val Ile Ile Val Ile

1608-  ATC ATC GTG CTT CTT TAT AGA CTC AGA AGG AGC ATG GGT AAT ATG CTA GGT AGC AGG AGG ATC CCA AGG GAC GAC TAC ACA TTA GAG  -546
       Ile Ile Val Leu Leu Tyr Arg Leu Arg Arg Ser Met Gly Asn Met Leu Gly Ser Arg Arg Ile Pro Arg Asp Asp Tyr Thr Leu Glu

1692-  CCG AAG ATC AGA CAT ATG TAC ACA AAC GGT TTT GAT GCA ATG GCT GAT GAG AAA AGA TGATCACGACCATTATCAGATGTCTTGTAAAGCAG
       Pro Lys Ile Arg His Met Tyr Thr Asn Gly Phe Asp Ala Met Ala Asp Glu Lys Arg ***

1784-  GCATGTATCCGTTGAGATCTGTATATAAT
```

24

1-AGGGATAAAGTCCCTTGTGAGTGCTTGATTGCAAAACTCTCCCCTTGGGAAAC ATG ACA GCA TAT ATC CAG AGA TCA CAG TGC ATC TCA ACA TCA -14
                                                        Met Thr Ala Tyr Ile Gln Arg Ser Gln Cys Ile Ser Thr Ser

96- CTA CTG GTT CTC ACC ACA TTG GTC TCG TGT CAG ATT CCC AGG GAT AGG CTC TCT AAC ATA GGG GTC ATA GTC GAT GAA GGG -42
     Leu Leu Val Leu Thr Thr Leu Val Ser Cys Gln Ile Pro Arg Asp Arg Leu Ser Asn Ile Gly Val Ile Val Asp Glu Gly

180- AAA TCA CTG AAG ATA GCT GGA TCC CAC GAA TCG AGG TAC ATA GTA CTG AGT CTA GTT CCG GGG GTA GAC TTT GAG AAT GGG TGC -70
     Lys Ser Leu Lys Ile Ala Gly Ser His Glu Ser Arg Tyr Ile Val Leu Ser Leu Val Pro Gly Val Asp Phe Glu Asn Gly Cys

264- GGA ACA GCC CAG GTT ATC CAG TAC AAG AGC CTA CTG AAC AGG CTG TTA ATC CCA TTA AGG GAT GCC TTA GAT CTT CAG GAG GCT -98
     Gly Thr Ala Gln Val Ile Gln Tyr Lys Ser Leu Leu Asn Arg Leu Leu Ile Pro Leu Arg Asp Ala Leu Asp Leu Gln Glu Ala

348- CTG ATA ACT GTC ACC AAT GAT ACG ACA CAA AAT GCC GGT GCT CCA CAG TCG AGA⟶TTC TTC GGT GTG ATT GGT ACT ATC GCA -126
     Leu Ile Thr Val Thr Asn Asp Thr Thr Gln Asn Ala Gly Ala Pro Gln Ser Arg Phe Phe Gly Ala Val Ile Gly Thr Ile Ala

432- CTT GGA GTG GCG ACA TCA GCA CAA ATC ACC GCA GGG ATT GCA CTA GCC CTA TTA CTA GGA GAG GCG AAA GAC ATA GCG CTC ATC -154
     Leu Gly Val Ala Thr Ser Ala Gln Ile Thr Ala Gly Ile Ala Leu Ala Leu Leu Leu Gly Glu Ala Lys Asp Ile Ala Leu Ile

516- AAA GAA TCG ATG ACA AAA ACA CAC AAG TCT ATA GAA CTG CTG CAA AAC GCT GTG GGG GAA CAA ATT CTT GCT CTA AAG ACA CTC -182
     Lys Glu Ser Met Thr Lys Thr His Lys Ser Ile Glu Leu Leu Gln Asn Ala Val Gly Glu Gln Ile Leu Ala Leu Lys Thr Leu

600- CAG GAT TTC GTG AAT GAT GAG ATC AAA CCC GCA ATA AGC GAA TTA GGC TGT GAG GGC ATA GGT TTA AGA CTG ATA AAA TTG -210
     Gln Asp Phe Val Asn Asp Glu Ile Lys Pro Ala Ile Ser Glu Leu Gly Cys Glu Gly Ile Gly Leu Arg Leu Ile Lys Leu

684- ACA CAG CAT TAC TCC GAG CTG TTA ACT GCG TTC GGC TCG AAT TTC GGA ACC ATC GGA GAG AAG AGC CTC ACG CTG CAG GCG CTG -238
     Thr Gln His Tyr Ser Glu Leu Leu Thr Ala Phe Gly Ser Asn Phe Gly Thr Ile Gly Glu Lys Ser Leu Thr Leu Gln Ala Leu

768- TCT TCA CTT TAC TCT GCT AAC ATT ACT GAG ATT ATG ACC ACA GGG CAG TCT AAC ATC TAT GAT GTC ATT TAT ACA -266
     Ser Ser Leu Tyr Ser Ala Asn Ile Thr Glu Ile Met Thr Thr Gly Gln Ser Asn Ile Tyr Asp Val Ile Tyr Thr

852- GAA CAG ATC AAA GGA ACG GTG ATA GAT GTG ATA GAT CTA GAG AGA TAC GTC ACC CTG TCT GTC AAG ATC CCT ATT CTT TCT GAA -294
     Glu Gln Ile Lys Gly Thr Val Ile Asp Val Ile Asp Leu Glu Arg Tyr Val Thr Leu Ser Val Lys Ile Pro Ile Leu Ser Glu

FIG. 1. Nucleotide sequence of the cDNA for F_0 mRNA and the predicted protein sequence. The hydrophobic amino-terminal and carboxy-terminal domains are underlined. The fusogenic domain is marked with a double underline. The arrow shows the predicted site of proteolytic cleavage. The positions of cysteine residues (circles) and potential asparagine-linked acceptor sites for carbohydrate (boxes) are indicated.[37]

25

The HL activity of nonaged virions is weak, whereas that of aged ones is strong. This increase in HL activity with aging seems to depend on the formation of large IMPs. In aged virions, large bundles of hydrophobic N-terminal fusogenic domains are produced by the aggregation of the membrane domains of F_1-forming spikes on the envelopes, causing hemolysis of erythrocyte membranes. The same phenomenon may explain the increase in HL activity of HVJ on sonication.[45]

Experimental System for Cell Fusion Induced by Sendai Virus

Sendai virus (HVJ) can fuse cells with viral receptors on their surface, including almost all cultured cells derived from mammals and fowls. In general, cells of established lines and transformed cells are fused more readily than differentiated cells,[46] and HVJ can fuse both cells in suspension and plated cells in culture dishes. The use of cells in suspension is easier for quantitative estimation of cell fusion efficiency. Here we describe the mechanism of fusion of Ehrlich ascites tumor (EAT) cells induced by HVJ.

Standard Procedure

The standard procedure for EAT cell fusion by HVJ is as follows: various concentrations of partially purified HVJ and EAT cells (final concentration, 1×10^7 cells/ml) are mixed in 1 ml of balanced salt solution (BSS: 140 mM NaCl, 5.4 mM KCl, 0.34 mM Na_2PO_4, 0.44 mM KH_2PO_4, buffered with 10 mM Tris-HCl at pH 7.6) containing 0.5 to 1 mM $CaCl_2$. Serum is a strong inhibitor of cell fusion. The tube is first incubated in an ice bath for 5 min to allow maximum adsorption of virus particles to the cells, and then incubated in a water bath at 37° under aerobic conditions with shaking. With this procedure, the fusion reaction with HVJ begins simultaneously in all the cells. After incubation for 30 min, spherical fused cells can be seen. The efficiency of cell fusion is calculated as the fusion index (FI):

$$FI = \frac{\text{Number of cells in control without virus}}{\text{Number of cells treated with virus}} - 1.0 \qquad (1)$$

Optimum Virus Concentration

The adsorption of multiple virus particles by cells is necessary for cell fusion.[47] A minimum of 200 virus particles per cell is required for fusion of

---

[45] Y. Hosaka, *Biken J.* **1,** 70 (1958).
[46] Y. Okada and J. Tadokoro, *Exp. Cell Res.* **32,** 417 (1963).
[47] Y. Okada and F. Murayama, *Exp. Cell Res.* **52,** 34 (1968).

EAT cells. The number required seems to depend on the type of cell, and for fusion of KB cells about 100 virus particles per cell are needed. Cell fusion efficiency increases with an increase in virus concentration, but the relationship of the efficiency of fusion with the viral concentration varies, depending on the condition of the cells (see below). As an optimum concentration for analysis of EAT cell fusion, about 500 to 1500 HAU of HVJ is generally used with 1×10^7 EAT cells. This corresponds to the addition of 1200 to 3600 virus particles per cell.

Optimum pH Range

A neutral pH range is optimal for cell fusion by HVJ,[60] in contrast to the low pH range required for cell fusion by Semliki Forest virus (SFV; Togaviridae), vesicular stomatitis virus (VSV; Rhabdoviridae), or influenzavirus (Orthomyxoviridae). This is due to the difference in the functional pH range of the fusogenic glycoproteins of these viruses.[48] Hemolytic activity is a functional marker of the fusogenic activity. Hemolysis by HVJ is detectable in a neutral pH range and decreases in the acidic range. The decrease of hemolysis (HL) at low pH is not as marked as the decrease in cell fusion: at pH 6, no cell fusion occurs, as shown in Fig. 2, but HL decreases only about 10 to 30% from that at neutrality. This discrepancy is due to the difference in the mechanisms of HL and cell fusion, but not in the function of the F protein itself. F proteins actually react with EAT cell membranes at pH 6, which results in cell lysis, but not cell fusion (Fig. 2). As described later, many factors are needed for the progression of the fusion reaction of EAT cells, after the interaction of the F protein with the cell membrane.

Virus – Cell Interaction at Low Temperatures

Fusion of cells occurs via fusion of the lipid bilayers of the plasma membranes. This step is a physiological reaction decreasing the potential energies of the lipid bilayers under conditions in which they have a fluid nature. At temperatures below 15°, the lipid bilayers do not have a fluid nature, and so the fusion reaction does not occur. In this section, the virus – cell interaction observed at 0° is described. No cell – cell fusion or viral envelope – cell membrane fusion occurs at 0°, but some interactions that are steps in cell fusion at 37° can be seen.

Insertion of N terminus of F_1 into Plasma Membrane

Virus – cell interaction at 0° shows a decrease in the transmembrane potential.[49,50] The potential of the cells, which is normally $-18.2 \pm$

[48] I. A. Wilson, J. J. Skehel, and D. C. Wiley, *Nature (London)* **289**, 366 (1981).

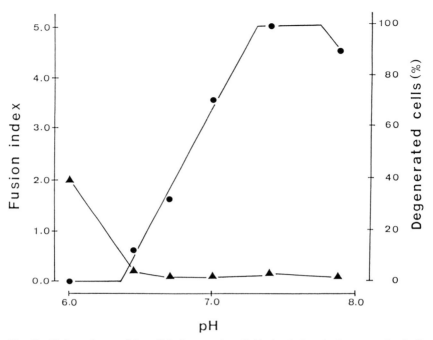

FIG. 2. pH dependence of the cell fusion reaction. ●, Fusion index; ▲, degenerated cells.[60]

2.1 mV, decreases to -6.1 ± 2.5 mV, and the membrane resistance also decreases from 491 to 162 Ω cm². These changes are not observed when the virions contain inactive F_0. This partial damage of the cell membranes may be due to the insertion of the N terminus of F_1. The insertion of the terminal sequence of F_1 into the lipid bilayer of human erythrocytes at $0°$ has been demonstrated directly by photoaffinity labeling.[51]

Formation of Huge Cell Aggregates

The first attachment of virus particles to the cell membrane is due to the binding of HANA to the terminal sialosaccharide residues of glycoproteins distributed on the cell surface. After this binding, there is still some distance between the N terminus of F_1 of the virions and the lipid bilayer of

[49] Y. Okada, *in* "Membrane Research" (F. Fox, ed.), p. 371. Academic Press, New York, 1972.
[50] Y. Okada, I. Koseki, J. Kim, Y. Maeda, T. Hashimoto, Y. Kanno, and Y. Matsui, *Exp. Cell Res.* **93**, 368 (1975).
[51] A. Asano and K. Asano, *Tumor Res.* **19**, 1 (1984).

the cell membrane. For insertion to take place, virions must come closer to the cell membrane. This is achieved by HANA – receptor interaction, that is, by stepwise binding of HANA proteins to sialo residues that have strong binding activities like those of sialolipids. Virions come in closer contact with the lipid bilayer, allowing insertion of the N terminus of F_1. The binding of HANA with receptors on the cells should strengthen the binding of the virus to the cells. This strong binding results in the appearance of huge cell aggregates on incubation at $0°$. The aggregation of EAT cells induced by influenzavirus is much less than that induced by HVJ in the neutral pH range.

Close Approximation of Plasma Membranes

Immediately after viral adsorption, EAT cell aggregates show hemadsorption activity, but later the virus particles adsorbed on the cells in huge aggregates are engulfed by the cell membrane. This event is associated with the disappearance of hemadsorption activity.[50] As a result, the cell membranes of adjacent cells in the aggregates come in close contact, bridged by the virions (Fig. 3), and fusion of the two cell membranes seems to proceed at $37°$ at these locations.

Virus – Cell Interaction at $37°$

Cell surfaces are covered by hydrophilic molecules, including saccharides, which form a barrier against fusion between cells. The most important factor for fusion is the exposure of naked lipid bilayers of the plasma membranes without these hydrophilic molecules. This exposure may permit direct contact of adjacent lipid bilayers by hydrophobic forces, and should be followed by their fusion. These reactions are observed during the interaction of HVJ and EAT cells at $37°$.

When incubated at $37°$, the huge cell aggregates that have been formed at $0°$ dissociate from the initial cell aggregates and the cell fusion reaction in the aggregates proceeds concomitantly. Thus, in single-cell suspensions, such as those of EAT cells, the extent of cell fusion depends on the balance between the rate of cell dissociation and the rate of cell fusion. At $37°$ cell dissociation is rapid, about 80% of the cells dissociating from the cell aggregates in the first 5 min.[52] At this temperature cell – cell fusion must occur before the cells have time to dissociate. There is some evidence that virus – cell interaction occurs within 5 min on incubation at $37°$.

[52] Y. Okada, F. Murayama, and Y. Yamada, *Virology* **27**, 115 (1966).

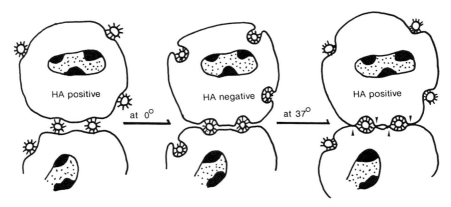

FIG. 3. Model of cell–virus interactions at 0 and 37°. Cell–cell fusion is allowed at direct sites of contact of the two plasma membranes (indicated by arrowheads), which appear to be different from the sites of fusion of viral envelopes with the cell membranes. HA, Hemagglutinin.

Breakdown of Cell Membrane Structure as Ion Barrier

On incubation at 37°, the potential across the cell membrane promptly decreases to about 0 mV and then rapidly returns to the normal level in about 30 min (Fig. 4).[50] This change is not observed when the virions contain inactive F_0. Thus, at 37° the first decrease may be due to the trapping of cholesterol by the N-terminal structure of F_1, which has been inserted into the lipid bilayer of the cell membranes at 0°. As a result, phase transition, or micelle formation, may occur in the lipid bilayer at the sites where virions are adsorbed, and minipores may be formed in the cell membrane. These pores are then sealed, and the potential is increased by the action of plasma membrane ion pumps. Calcium ions may be required for sealing, because in medium depleted of calcium ions the virus induces only cell lysis, with no cell fusion.

Evidence for the formation of pores and their sealing is the finding that diphtheria toxin fragment A (22 kDa) can penetrate through the cell membranes only at the stage of low potential, and not later during incubation at 37°, and that it cannot penetrate through the membrane during cell–virus interaction at 0°.[53] The transient formation of pores in the membrane is used to introduce macromolecules into cells in order to examine their effects on the cells. For instance, as seen in Fig. 5, the introduction of UV-specific endonuclease V (16 kDa), derived from T4 bacteriophage, into the cytoplasm of xeroderma pigmentosum (XP) cells shows that it is functional in human cells, and can rescue the defective

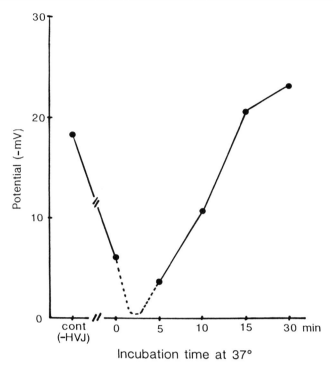

FIG. 4. Alteration in potential across the cell membrane during the cell fusion reaction. A mixture of 10^7 FL cells in suspension and 750 HAU of HVJ is incubated at 0° for 5 min and then at 37° for 0–30 min. The time for the lowest potential is estimated at about 2 min, based on observation of the change in the form of mitochondria. (Modified from Okada et al.[50])

unscheduled DNA synthesis of these cells. The introduction procedure is simple: a mixture of UV-inactivated HVJ and the protein is added to monolayer cultures of XP cells, and after incubation for a few minutes at 0°, the cultures are then incubated at 37° for 15 to 20 min.[54]

Transformation of Mitochondria

Changes in the shape of mitochondria in cells occur at the time of low potential. During incubation at 37°, change from an "orthodox" to a "condensed" configuration is detectable after 2 min; but within 60 min the mitochondria return to their "orthodox" form.[55] The condensed configu-

[53] M. Yamaizumi, T. Uchida, and Y. Okada, *Virology* **95**, 218 (1979).
[54] K. Tanaka, M. Sekiguchi, and Y. Okada, *Proc. Natl. Acad. Sci. U.S.A.* **72**, 4071 (1975).
[55] J. Kim and Y. Okada, *Exp. Cell Res.* **130**, 191 (1980).

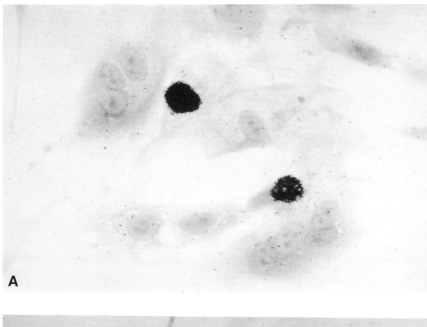

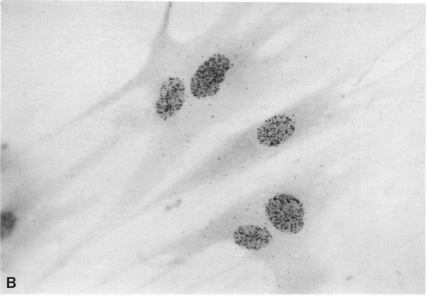

FIG. 5. Rescue of defective, unscheduled DNA synthesis of xeroderma pigmentosum cells by introduction of UV-specific endonuclease V derived from T4 bacteriophage (B). This enzyme is introduced by the addition of HVJ. (A) Control without HVJ.

ration resembles that which is generally observed when mitochondria are isolated from cells or tissues. Isolated mitochondria are known to contain an average of 0.75 mM calcium ions per kilogram wet tissue. This level is significantly higher than that in mitochondria *in vivo* and is due to the accumulation of calcium ions in the mitochondria during the isolation procedure.[56]

When the ion barrier breaks down, cells must control the high concentration of calcium ions introduced into their cytoplasm from outside the cells. The gradient of calcium concentration across the cell membrane is on the order of 10^4, and a high intracellular calcium ion concentration could result in cell death. In fact, the concentration of free calcium ions in the cytoplasm increases no more than a few fold over the physiological level during the cell fusion reaction (Y. Okada, unpublished data, 1992). Mitochondria are known to control the level of calcium ions introduced into the cytoplasm on cell injury, and their capacity to trap calcium ions is higher than that of other organelles: about 40 to 45% of the intracellular calcium ions is found in mitochondria.[56] Thus, the observed change in the form of mitochondria during fusion is probably due to their action in rapidly trapping calcium ions. This buffering action of mitochondria seems to be correlated with the requirement for an energy supply at the time of the cell fusion reaction.

Appearance of Stage Requiring Calcium Ions and Energy Generation

When calcium ions are removed from the reaction medium, cell lysis becomes predominant and cell fusion is suppressed.[57] The stage requiring calcium ions occurs immediately after the start of incubation at 37°, which overlaps the stage of low potential. After incubation for about 10 min at 37°, calcium ions are no longer required (Fig. 6). Calcium ions can be replaced by strontium, barium, or manganese ions, but not by magnesium ions, which inhibit the action of calcium ions. Similar effects of divalent cations are observed during mechanical fusion induced by electroporation.[58]

The role of calcium ions in the cell fusion reaction is unknown, but possible actions are as follows: (1) sealing of the minipores in the lipid bilayer of plasma membranes, induced by the fusogenic domain (calcium ions inhibit the HL activity of the virus), (2) dissociation of the binding between intramembrane particles (IMPs) and the inner cytoskeleton sys-

[56] G. Fiskum and A. L. Lehninger, *in* "Calcium and Cell Function" (W. Y. Cheung, ed.), Vol. 2, p. 39. Academic Press, New York, 1982.
[57] Y. Okada and F. Murayama, *Exp. Cell Res.* **44,** 527 (1966).
[58] T. Ohno-Shosaku and Y. Okada, *J. Membr. Biol.* **85,** 269 (1985).

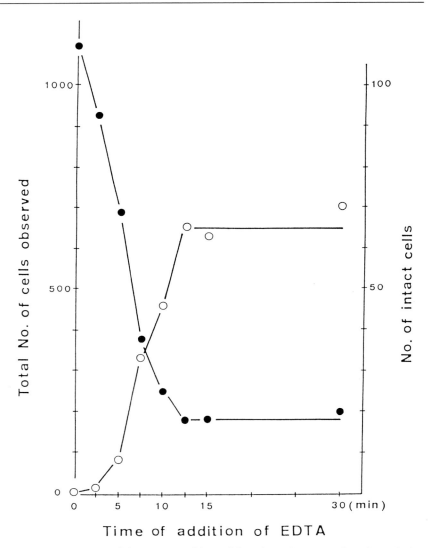

FIG. 6. Appearance of the stage requiring calcium ions. At appropriate times during incubation of the EAT cell–HVJ complex at 37°, calcium ions are removed by the addition of ethylenediaminetetraacetic acid (EDTA) and, after incubation for 60 min at 37°, the cells are observed. ●, Total number of cells observed; ○, number of intact cells, including fused cells.[57]

tem by the entry of calcium ions into the cytoplasm, and (3) participation in direct attachment of the two lipid bilayers and the removal of water bound to surface molecules of the plasma membranes.

The stage of calcium ion requirement overlaps the stage of energy requirement during cell fusion. The addition of inhibitors of oxidative phosphorylation, such as 2,4-dinitrophenol, NaN_3, or NaCN, to the reaction medium inhibits cell fusion and results in cell lysis.[59,60] The further addition of glucose to the medium inhibits cell lysis induced by these reagents and promotes cell fusion.[61] These results show that an ATP-generating system in the cells supports cell fusion, and that suppression of ATP generation inhibits cell fusion and induces cell lysis. No lysis occurs with virions containing inactive F_0. This suggests that the formation of minipores in the lipid bilayer of plasma membranes by the action of F_1 is the factor responsible for the appearance of this requirement for energy. The end of the stage requiring energy coincides with the end of the stage requiring calcium ions.

What kind of organelle requires energy and for what function? Energy is required most probably by mitochondria for trapping excess free calcium ions introduced into the cytoplasm from outside the cells. The cells seem to be protected from the lethal effect of excess calcium ions by this function of mitochondria. A preliminary experiment shows that when a cell–virus mixture is incubated with 5 μM NaN_3 at 37°, a rapid increase in intracellular calcium ions begins after about 30 sec, in association with cell lysis (Y. Okada, unpublished observation, 1992). There are reports that (1) calcium ions enter mitochondria electrophoretically, due to the membrane potential (inside negative) across the inner membrane induced by electron transport; and (2) when respiration is interrupted by inhibitors such as cyanide, mitochondria can still accumulate and retain calcium ions provided ATP is available, because ATP hydrolysis by mitochondria also generates a transmembrane potential and thus can support calcium ion uptake.[56] These findings can explain the requirement for energy observed in the cell fusion reaction.

The fusion of EAT cells is reported to be affected by the preincubation conditions of the cells at 37°, before the addition of HVJ.[52] Fusion is markedly enhanced by preincubation under aerobic conditions, and decreased by preincubation under anaerobic conditions. These changes, which are reversible and can be repeated several times as shown in Fig. 7, correspond to changes in the ATP level, depending on the preincubation

[59] Y. Okada, *Exp. Cell Res.* **26**, 98 (1962).
[60] Y. Okada, *Exp. Cell Res.* **26**, 119 (1962).
[61] A. Yanovsky and A. Loyter, *J. Biol. Chem.* **247**, 4021 (1972).

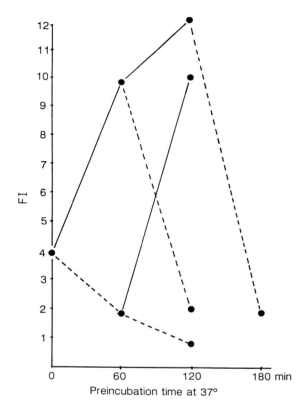

FIG. 7. Alteration of fusion capacity of EAT cells induced by preincubation at 37° under aerobic or anaerobic conditions. Test tubes containing 0.5 ml (1.2 × 10⁷ cells) are preincubated at 37° with (solid line) or without (dashed line) shaking for the indicated length of time, before the cell fusion reaction with 4000 HAU of HVJ under standard conditions.[52]

conditions. The extent of fusion of EAT cells seems to depend on the calcium buffering capacity of the mitochondria at the beginning of the cell fusion reaction at 37°. If buffering by mitochondria is slow, the rate of cell–cell fusion will be slow, and the balance between disaggregation of cells and cell fusion will shift to disaggregation, resulting in low efficiency of cell fusion.

Induction of Mobility in Intramembrane Particles, Appearance of Direct Attachment of Plasma Membranes, and Progress of Cell–Cell Fusion

At the time of low potential, various other organelles besides mitochondria transiently change in configuration, and later regain their original configurations.[55] The changes observed are distension of the endoplasmic reticulum (ER) and Golgi stacks and enlargement of the inner space of the nuclear envelopes. The arrangement of 100-Å filaments also becomes disordered. These changes resemble those generally observed on isolation of these organelles, and may be due to the disorder of the cytoskeleton system, independent of the change in the structure of the organelles induced by the introduction of calcium ions into the cells.

The plasma membrane structure of the cells is also modified transiently at this time. Examination of freeze-fractured plasma membranes shows evidence of "cold-induced clustering" of IMPs on the P fracture face.[62] Their appearance suggests that IMPs become mobile in the lipid bilayer of the plasma membrane, possibly due to their dissociation from the cytoskeleton system under the membrane, caused in turn by an increase in intracellular calcium ions. This modification permits direct attachment of adjacent plasma membranes at the sites formed at $0°$ (Fig. 3), which can be seen in thin-section electron micrographs. The rapid fusion of cells may occur through these sites, which seem to be distinct from sites of viral envelope–cell membrane fusion.[63]

Inhibition of Cell–Cell Fusion by Cytochalasin D

Fusion of EAT cells by HVJ is completely inhibited by cytochalasin D (CD) at 5 μg/ml. Virus–cell interaction at $0°$ proceeds both in the presence or absence of CD: virions are adsorbed onto the cells and the cells aggregate. The virions are then engulfed by cell membranes, resulting in the disappearance of hemadsorption activity on the cell surface. When the cell–virus complex is incubated at $37°$, these initial reactions can also proceed as in the absence of CD. The hemadsorption activity reappears on the cell surfaces, the viral envelopes fuse with cell membranes to the same extent, with and without CD, and the stages that are sensitive to sodium azide and require calcium ions occur. But cell–cell fusion is completely inhibited and the cell aggregates dissociate into single cells. This observation indicates that cell–cell fusion is distinct from viral envelope–cell membrane fusion in this type of cell fusion.[64]

These findings suggest that the fusogenic domain at the N terminus of F_1 interacts with the cell membrane in the presence and absence of CD. The reason cell–cell fusion is inhibited under these conditions is explained by the finding that no cold-induced clustering of IMPs is detectable in cells

[62] J. Kim and Y. Okada, *Exp. Cell Res.* **132,** 125 (1981).
[63] Y. Okada, *in* "Current Topics in Membranes and Transport: Membrane Fusion in Fertilization, Cellular Transport and Viral Infection" (N. Düzgüneş and F. Bronner, eds.), Vol. 32, p. 297. Academic Press, San Diego. 1988.
[64] Y. Miyake, J. Kim, and Y. Okada, *Exp. Cell Res.* **116,** 167 (1978).

treated with CD, at the stage showing strong IMP clustering in the absence of CD.[62] This suggests that CD inhibits the action of calcium ions in causing dissociation of IMPs from the cytoskeleton system under the membrane. Cytochalasin D has a strong affinity for microfilaments, but it is not clear whether the CD–microfilament complex inhibits this dissociation, or whether native microfilaments themselves participate in this kind of cell fusion.

Fusion of Liposomes with Naked Area Appearing on Plasma Membranes Induced in Cell Fusion Reaction

Cells infected with subacute sclerosing panencephalitis (SSPE) virus are selectively killed by liposomes containing diphtheria toxin fragment A.[65] The SSPE virus is a mutant of measles virus that has a defect in budding at the maturation step, due to the dysfunction of the M protein. Thus, cells persistently infected with the virus do not produce infectious cell-free virus, and they can be successfully cultured and passaged only by cell–cell infection during cocultivation with uninfected cells. When SSPE virus-infected cells (SSPE cells) are sparsely seeded onto monolayer cultures of normal cells, infection spreads by cell fusion between the SSPE cells and the native cells. As a result, giant cells grow in size and then autolyse, leaving vacant sites, like plaques, in the cell sheet. As long as SSPE cells remain in the culture, giant cells continue to form in the regions around vacant areas, and thus eventually the whole native cell sheet is destroyed. When liposomes containing fragment A are added to the culture plate, however, the cell sheets are cured and appear similar to sheets of uninfected cells after 72 hr, because the uninfected cells grow out over the vacant areas after the liposomes have killed the SSPE cells and disrupted the giant cells completely. The liposomes do not affect the uninfected cells in the cultures. These observations may be explained by the fact that some naked regions without IMPs appear on the cells at the stage preceding the cell fusion reaction as described, and by supposing that liposomes can fuse selectively with these naked regions of the cells.

Fusion of Viral Envelopes with Cell Membranes

As described previously, viral envelope fusion is essential for viral infection, and the viral glycoproteins F and HANA are integrated into the plasma membranes of fused EAT cells. Envelope fusion seems to be completed after cell–cell fusion, and proceeds even when cell–cell fusion is inhibited in the presence of cytochalasin D.[63] The initial reaction of this

---

[65] S. Ueda, T. Uchida, and Y. Okada, *Exp. Cell Res.* **132**, 259 (1981).

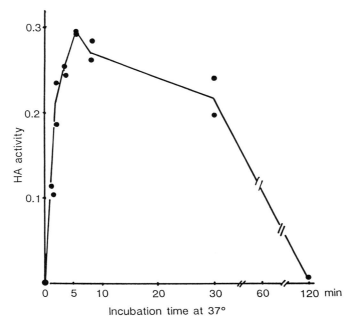

FIG. 8. Appearance of HA activity on cell aggregates during incubation at 37° and its decrease on further incubation. Incubation of the cell–virus mixture at 0° leads to the disappearance of the HA activity of the cell aggregates. At 37°, HA activity reappears promptly and then decreases gradually. Hemadsorption activity is expressed as OD_{540} values of human erythrocytes adsorbed maximally on the cell aggregates, after hemolysis with saponin.[50]

envelope fusion is also induced by the fusogenic domain at the N terminus of F_1, but the mechanisms of fusion of the viral envelope with the cell membrane are not understood well.

Figure 8 shows a kinetic curve of the reappearance of hemadsorption (HA) activity of the cell–virus complex and its decrease on incubation at 37°. On incubation at 37° the HA activity reappears promptly, reaches a maximum after 5 min, and then decreases gradually.[49] After incubation for 2.5 min, when cell–cell fusion is occurring, no viral envelope fusion can be observed. The decrease in HA activity from a plateau is not due to liberation of virus particles from the cell surface, but to internalization of viral glycoproteins integrated into the cell membrane through coated vesicles.[66] Thus, envelope fusion may occur after incubation for about 5 min, when the HA activity reaches a plateau.

[66] J. Kim and Y. Okada, *Exp. Cell Res.* **140,** 127 (1975).

Fusion of Cultured Cells Plated on Dishes

Subconfluent cultures can be used for cell fusion. The culture medium is removed and the cells are washed once with BSS containing 1 mM CaCl$_2$, and treated with a suspension of HVJ in BSS containing calcium ions. The optimal concentration of HVJ depends on the characteristics of the cells. After incubation for 5 min under ice-cold conditions to allow maximum adsorption of HVJ, the dishes are incubated at 37° for 30 min. For hybrid cell formation, cocultures of two parent cells are used, and the cells are fused with inactivated HVJ (see the next section).

Plated cells do not show mutual cell repulsion, unlike cells in suspension; therefore, fast fusion to overcome strong mutual repulsion is unnecessary, unlike the case of EAT cells. In the fusion of erythrocyte ghosts, which show weak mutual repulsion, slow fusion can be achieved through viral envelope fusion with two erythrocyte ghosts.[67] In this case, calcium ions and energy generation do not seem to be essential. It is difficult to explain which pathway is dominant in fusion of cells plated on dishes. The area essential for fusion of adjacent cells is small, compared with the total area of the cell surface available for adsorption of virus; thus, the probability of virus adsorption to this area is very low. Almost all virus particles may be adsorbed to the apical area rather than the sites of fusion. In fact, the site of adsorption of virus to monolayer cultures of MDCK cells (apical side) was found to be different from the adsorption site required for cell fusion (lateral side).[68]

Inactivation of Sendai Virus Genome

In some cases, such as in hybrid cell formation, viral growth in fused cells must be prevented by inactivation of HVJ. Two methods are available for HVJ inactivation. One is inactivation by UV irradiation: Uric acid must be removed from virus stock by dialysis, or semipurification by differential centrifugation. Irradiation under a UV lamp at 100 erg/sec/mm^2 for 2 min is sufficient for inactivation of the infectivity of 1 ml of $1-5 \times 10^4$ HAU of virus in a Falcon (Becton Dickinson, Oxnard, CA) dish of 3-cm diameter. Usually, the virus sample is irradiated for 4 min (twice the necessary time). The other method is inactivation by treatment with β-propiolactone.[69] One volume of a solution of 0.03 to 0.13% β-propiolactone, 1.68 g sodium bicarbonate, and 0.4% (v/v) phenol in 1000 ml of

[67] S. Knutton, *J. Cell Sci.* **28**, 189 (1977).

[68] J. Kim and Y. Okada, *Eur. J. Cell Biol.* **51**, 128 (1990).

[69] H. Koprowski, F. C. Johnson, and Z. Steplewski, *Proc. Natl. Acad. Sci. U.S.A.* **58**, 127 (1967).

saline is mixed with 9 vol of virus sample, and the mixture is incubated successively at room temperature for 10 min, at 37° for 2 hr, and overnight in a refrigerator. During the incubation the virus is inactivated and β-propiolactone is hydrolyzed completely. The cell fusion activity of HVJ is not affected by these inactivation procedures. Reactivation of growth of the virus in fused cells seems unlikely, and has never been observed experimentally. Moreover, as mentioned earlier, even if the virus were reactivated the virus progeny grown in cells would contain inactive F_0, and therefore would be noninfectious.

Summary and Discussion

As described in this chapter, EAT cells are rapidly fused by HVJ at 37°, under conditions supplying calcium ions and allowing energy generation. The initial reaction is disruption of the plasma membranes, induced by the function of the fusogenic N-terminal domain of F_1. Without calcium ions, cell fusion is inhibited and the cells die, probably because the disrupted membrane structure is not repaired. On the other hand, in the presence of calcium ions a high concentration of calcium ions is introduced into the cells transiently, and removal of these calcium ions is essential for preventing the death of the cells. These calcium ions are mainly taken up by mitochondria, and the requirement of energy generation seems to be primarily for the function of the mitochondria. On introduction of calcium ions, IMPs become mobile, due to their dissociation from the cytoskeletal system under the membrane. At the sites where the IMPs are removed direct contact of the lipid bilayers of the plasma membranes becomes possible, and fusion of these plasma membranes can occur.

For dissociation of the IMPs from the cytoskeletal system under the membrane, the concentration of calcium ions may not need to be as high as that introduced into the cells. In this chapter, observations in media containing 0.5 to 1 mM calcium ions are described, but the efficiencies of lower concentrations, such as 10 to 100 μM, should also be tested under various conditions. The HL activity of the virus increases with aging. The level of HL may correspond to the extent of destruction of the plasma membranes by F_1. A higher concentration of calcium ions seems to be needed for fusion of cells, when the HVJ has a higher HL activity.

Acknowledgments

The author wishes to thank Drs. S. Shigekawa (National Cardiovascular Center, Osaka), A. Asano (Institute for Protein Chemistry, Osaka University), and J. Kim (Kyoto College of Pharmacy) for helpful discussions. This chapter is dedicated to my colleague Dr. T. Uchida, who died on May 3, 1989.

[4] Kinetics of Cell Fusion Mediated by Viral Spike Glycoproteins

By STEPHEN J. MORRIS, JOSHUA ZIMMERBERG, DEBI P. SARKAR, and ROBERT BLUMENTHAL

It has been recognized for some time that cell–cell fusion can occur as a result of virus–cell interaction.[1] The fusion can be induced either after the virus fuses with the plasma membrane, or after biosynthesis and expression of the viral envelope protein on the cell surface. Assays for fusion of intact virus with cells have been described elsewhere in this series.[2,3] Here we shall deal with assays for the fusion activity of viral proteins expressed on the surface of nucleated cells.

Fusion of cells mediated by viral envelope proteins was first assessed by light microscopic observation (syncytium formation) and electron microscopic techniques,[1] as well as by determination of the transfer of biochemical markers from one cell to another.[4] To view the results of those assays, the cells are activated and then placed into physiological growth conditions for several hours. Although they provide a reasonable quantification of the extent of the fusion reaction, those assays are not designed to monitor kinetics that may provide information about initial steps of the fusion reaction. Using aqueous or lipid fluorophores, cell fusion assays have been performed to enumerate events in which fluorescence from a labeled membrane or the cytoplasm moves to an unlabeled membrane or cytoplasm.[5-10]

We have developed methods to monitor dye redistribution continuously during cell fusion mediated by viral envelope proteins. We measure fusion from the initial triggering of the reaction by monitoring the mixing of either membrane contents[11] or cytoplasmic contents, or both simulta-

---

[1] Y. Okada, *Curr. Top. Membr. Transp.* **32**, 297 (1988).

[2] D. Hoekstra and K. Klappe, this series, Vol. 220 [20].

[3] A. Puri, M. J. Clague, C. Schoch, and R. Blumenthal, this series, Vol. 220 [21].

[4] S. J. Doxsey, J. Sambrook, A. Helenius, and J. White, *J. Cell Biol.* **101**, 12 (1985).

[5] P. M. Keller, S. Person, and W. Snipes, *J. Cell Sci.* **28**, 167 (1977).

[6] J. W. Wojcieszyn, R. A. Schlegel, K. Lumley-Sapanski, and K. A. Jacobson, *J. Cell Biol.* **96**, 151 (1983).

[7] D. Hoekstra and K. Klappe, *Biosci. Rep.* **6**, 953 (1986).

[8] C. Kempf, M. R. Michel, U. Kohler, and H. Koblet, *Arch. Virol.* **95**, 283 (1987).

[9] Q. F. Ahkong, J. P. Desmazes, D. Georgrscauld, and J. A. Lucy, *J. Cell Sci.* **88**, 389 (1987).

[10] A. E. Sowers, *J. Cell Biol.* **102**, 1358 (1986).

[11] S. J. Morris, D. P. Sarkar, J. M. White, and R. Blumenthal, *J. Biol. Chem.* **264**, 3972 (1989).

neously.[12] Single cells are monitored by low light level fluorescence video microscopy and a population of cells by spectrofluorimetry. Because long-term survival of the cells is not required, experiments can be performed under defined conditions, and the effects of pH, osmolarity, temperature, metabolic inhibitors, pharmacological agents, toxins, and so on, on the fusion activity can be tested.

In this chapter we describe methods to monitor the fusion induced by the influenza virus hemagglutinin expressed on the surface of cells. Although this is a special case of cell–cell fusion, the methods are generally applicable to a variety of fusing systems.

Principle of Method

The kinetic assay is based on the increase in fluorescence due to dequenching of a dye in one cell as it fuses to a second cell, which allows dye dilution. This increase in fluorescence is critically dependent on both the dye concentration in the first, labeled cell, and the ratio of the size of the two cells. Too little dye will not be self-quenched. Too much dye will remain quenched even after dilution. The probe can either be hydrophobic and incorporated into the bilayer in self-quenching concentrations, or be hydrophilic and incorporated into the cytoplasm to be quenched by cytoplasmic components, for example, hemoglobin in the red blood cell (RBC). Self-quenching of soluble probes usually requires millimolar concentrations, which are difficult to achieve. The fluorescence self-quenching method was originally adapted from work with liposomes.[13] It has been applied extensively to virus–cell fusion (see other chapters in this volume[2,3]).

Erythrocytes are the cells of choice for spectrofluorimetric and video microscopic studies, because they may incorporate lipid and aqueous fluorophores. Membrane probes are self-quenched at high concentration; the fluorescence of the aqueous dyes is quenched by hemoglobin. However, studies with erythrocytes are limited to those virus strains (e.g., paramyxo- and orthomyxoviruses) that recognize sialoglycoproteins and lipids as their receptors on the target membrane. (Vesicular stomatitis virus, a rhabdovirus, will fuse with erythrocyte ghosts provided the erythrocyte membrane has a symmetrical distribution of lipid.[14]) Fusion of erythrocytes with cells

[12] D. P. Sarkar, S. J. Morris, O. Eidelman, J. Zimmerberg, and R. Blumenthal, *J. Cell Biol.* **109**, 113 (1989).
[13] J. N. Weinstein, S. Yoshikami, P. Henkart, R. Blumenthal, and W. A. Hagins, *Science* **195**, 489 (1977).
[14] S. Grimaldi, R. Verna, A. Puri, S. J. Morris, and R. Blumenthal, *Proc. Serono Symp.* **51**, 197 (1988).

induced by influenzavirus hemagglutinin has been measured, using the dequenching method, on a population of cells by spectrofluorimetry,[11,12] and with single cells using rapid-flow quantitative fluorescence microscopy.[15]

Spatial relocations of dyes associated with fusion can be monitored and quantified by video microscopy without the requirement of fluorescence dequenching. Therefore visualization methods can be applied to a broader range of cell fusion studies and can utilize probes that are not self-quenched. With nucleated cells probes may clear from the plasma membrane via lipid trafficking processes involved in normal cell maintenance.[16] However, lipid fluorophores such as octadecylrhodamine (R18) or 1,1-dioctadecyl-3,3,3',3'-tetra methyl indocarbocyanine perchlorate (DiI) remain on the plasma membrane of nucleated cells for > 24 hr (D. Dimitrov and R. Blumenthal, unpublished observations, 1990).

Labeling Erythrocytes with Soluble and/or Membrane Probes

Procedures for labeling erythrocytes consist of incubating the cells with appropriate dye solutions, followed by thorough washing to remove unincorporated label. Fluorescent probes for labeling membrane bilayers generally share the same structural properties as phospholipids: they are amphipathic molecules with low solubility in water. Because of the hydrophobicity, these dyes are usually dissolved at high concentration in an organic solvent such as dimethyl sulfoxide (DMSO), ethanol, or tetrahydrofuran, and then added to the suspensions of cells in small volumes with rapid mixing. Presumably this forms a short-lived supersaturated solution from which the insoluble molecules can equilibrate with the bilayer phase. The probe rapidly forms microcrystals as the supersaturation dissipates. Some membranes can be labeled by incubation with such crystals; presumably repeated collisions transfer a few molecules per encounter from the crystal into the bilayer. However, such long incubation times and constant stirring are detrimental to most biological materials.

Exact labeling conditions may require some empirical adjustments. For erythrocytes, lipid dyes such as R18[11,12] or DiI[15] remain in the plasma membrane and show little to no propensity to equilibrate with other unlabeled membranes, even if the two cells are bound through the viral hemagglutinin expressed on the cell surface. Detergents[17] or phospholipid

[15] D. Kaplan, J. Zimmerberg, A. Puri, D. P. Sarkar, and R. Blumenthal, *Exp. Cell Res.* **195,** 137 (1991).
[16] R. E. Pagano and R. G. Sleight, *Science* **229,** 1051 (1985).
[17] Z. Lojewska and L. M. Loew, *Biochim. Biophys. Acta* **899,** 104 (1987).